Business and Service Telephone Conversations

Business and Service Telephone Conversations

An Investigation of British English, German and Italian Encounters

Cecilia Varcasia
Research Fellow, Free University Bozen-Bolzano, Italy

© Cecilia Varcasia 2013

All rights reserved. No reproduction, copy or transmission of this publication may be made without written permission.

No portion of this publication may be reproduced, copied or transmitted save with written permission or in accordance with the provisions of the Copyright, Designs and Patents Act 1988, or under the terms of any licence permitting limited copying issued by the Copyright Licensing Agency, Saffron House, 6–10 Kirby Street, London EC1N 8TS.

Any person who does any unauthorized act in relation to this publication may be liable to criminal prosecution and civil claims for damages.

The author has asserted her right to be identified as the author of this work in accordance with the Copyright, Designs and Patents Act 1988.

First published 2013 by
PALGRAVE MACMILLAN

Palgrave Macmillan in the UK is an imprint of Macmillan Publishers Limited, registered in England, company number 785998, of Houndmills, Basingstoke, Hampshire RG21 6XS.

Palgrave Macmillan in the US is a division of St Martin's Press LLC,
175 Fifth Avenue, New York, NY 10010.

Palgrave Macmillan is the global academic imprint of the above companies and has companies and representatives throughout the world.

Palgrave® and Macmillan® are registered trademarks in the United States, the United Kingdom, Europe and other countries.

ISBN 978–1–137–28617–8

This book is printed on paper suitable for recycling and made from fully managed and sustained forest sources. Logging, pulping and manufacturing processes are expected to conform to the environmental regulations of the country of origin.

A catalogue record for this book is available from the British Library.

A catalog record for this book is available from the Library of Congress.

Typeset by MPS Limited, Chennai, India.

Contents

List of Tables and Figures vi
Acknowledgements vii
List of Abbreviations viii
Transcription Convention System ix

Introduction 1
1 Theoretical Framework 7
2 Data and Methodology 29
3 Simple Response Format to the Request 39
4 Response plus Extension 45
5 Insertion Sequence Followed by the Response 71
6 The Caller Leads the Conversation 96
7 The Different Response Formats at One Glance 108
8 Service Encounters and Call Centre Training Implications 118
9 Conclusions and Implications 138

Appendix 146
References 166
Subject Index 174
Author Index 180

List of Tables and Figures

Tables

2.1	Services called in the data	31
3.1	Simple response format	44
4.1	Extensions of the response	67
4.2	Types of increments of the response: the entire corpus	68
4.3	Types of increments of the response: cross-cultural overview	68
5.1	Actions performed by the insertion sequence	90
5.2	Actions in the response after the occurrence of an insertion sequence	91
5.3	Increment types in response format: insertion sequence + response	93
5.4	Type of response following the insertion sequence	94
6.1	R responses to C's solicitation of more talk	105
6.2	Actions in the delayed response expansion	105
6.3	Types of increments in solicited expansions of the response	107
7.1	Actions in the response: the corpus overall	110
7.2	Actions in the response: cross-cultural comparison	111
7.3	Free constituents and extension types of the increment of the response	115

Figure

7.1	Simple vs complex response formats in English, German and Italian	113

Acknowledgements

I am grateful to many people who have contributed to the research and supported my work at various stages. I am indebted the most to Gabriele Pallotti, who encouraged me to begin my study of telephone talk and has guided me from the very beginning. I am also very grateful to Greg Myers for his support and patient examination of previous drafts of this work. My appreciation and thanks also go to Paul Seedhouse, Hugo Bowles and two anonymous reviewers for their sometimes challenging comments and suggestions at various stages of the research.

I would also like to especially thank all the services who participated in the project and gave their consent for recording in the three countries. Many thanks to the people who helped me in the collection of the data and their transcription: Paola Contu, Gloria Deriu, Silvia Nieddu, Anna Tilocca, but also David, Steve, Tania, Juliane, Nina, Susan and Dario.

Last but not least, special thanks to Rosmarie de Monte Frick for her patient and prompt final linguistic review of the manuscript. Any mistakes or imperfections are entirely my responsibility.

List of Abbreviations

AP	adjacency pair
C	caller
CA	conversation analysis
DA	discourse analysis
NP	noun phrase
PCP	possible completion point
PSE	public service encounter
R	receiver
SFL	systemic functional linguistics
TCU	turn constructional unit
TRP	transition relevance place
Unattached NP	unattached noun phrase

Transcription Convention System

Below are the main conventions for transcription developed by Gail Jefferson and reported in more detail in Atkinson and Heritage (1984), ten Have (1999), Hutchby and Wooffitt (1998), Silverman (1998), Charles Antaki's online introduction to transcription http://www-staff.lboro.ac.uk/~ssca1/transintro1.htm, and Schegloff's online tutorial http://www.sscnet.ucla.edu/soc/faculty/schegloff/TranscriptionProject/index.html.

C	caller
R	receiver
.	marks a falling pitch or intonation
,	indicates a continuing intonation with slight upward or downward contour
?	indicates a rising vocal pitch or intonation
Italics	emphasis
CAPITAL	raised voice
lo::ng	stretched sounds
°quiet°	words spoken in a low voice
>speed-up<	increased speed of delivery
<speed-down>	decreased speed of delivery
[ ]	brackets indicate overlapping utterances
=	equal marks indicate contiguous utterances, or continuation of the same utterance to the next line
.hh/hh.	audible inbreath/outbreath.
ye(hh)s	h's within parentheses indicate within-speech aspiration, possibly laughter
(.)	full stop within parentheses indicates micropause
(2.0)	number within parentheses indicates pause of length in approximate seconds
((cough))	items within double parentheses indicate some sound or feature of the talk which is not easily transcribable, e.g. '((in falsetto))' and transcriber's comments on some extra-linguistic phenomenon
(yes)	parentheses indicate transcriber's doubt about hearing of passage
→ / **bold**	analyst's signal of a significant line

Introduction

The relationship between service providers and customers in service encounters is quite a delicate matter. The success of the business may depend on such relationships and the care taken of the customers. This study aims at investigating service encounters on the telephone in English, German and Italian from a conversation analysis (CA) perspective. Since the beginnings of CA, telephone calls have been widely researched. At the beginning, studies aimed at discovering regularities and patterns in this distinct type of conversation, above all in its opening and closing sequences. One strand of this research has tried to find common and different patterns among different languages through cross-cultural studies. Less attention has been paid to the development of these exchanges after the openings. This is the object of this book. Previous studies focused on the construction of such relationships between service providers and customers in face-to-face encounters, as in the work of the Pixi project in which English and Italian bookshop encounters were compared (Aston, 1988a; Zorzi, 1990; Brodine, 1991), and the interactions between the employees of a reprographic store's drop-off counter and their customers by Vinkhuyzen and Szymanski (2005), in which non-granting requests were analysed. The way speakers deal with the business on the phone has only recently been explored by Lee (2011a), in her analysis of a distinctive type of telephone conversation, i.e. that made to an airline service in South Korea. In her study she examines how the airline agents design their conduct when customers' requests cannot be satisfied and how they shape their conduct in a conciliatory direction. Most of the studies on telephone talk in service encounters have focused rather on the opening sequence of the conversation in different languages, as here speakers follow specific conversational routines and rituals (cf. Schegloff,

1986). Studies have considered phone calls in Finnish (Halmari, 1993), Finnish and German (Lieflander-Koistinen, 1992), Italian (Bercelli and Pallotti, 2002), Italian and English (Bowles and Pallotti, 2004; Varcasia, 2006), Italian and Spanish (Colamussi and Pallotti, 2003), Italian and German (Varcasia, 2003a; Thüne, 2003), and a comparison of openings in English, French, German, Italian and Spanish (Pallotti and Varcasia, 2008). My previous interest and study of telephone openings in service encounters (Varcasia 2003a, b, 2006, 2007, 2008; Pallotti and Varcasia, 2008) within a research project at the University of Sassari, Italy, led to an interest in looking at other sequential patterns in the call, such as the core business of the conversation. The focus of this book is on the responses provided by receivers of three European languages, namely British English, German and Italian, to callers' pre-requests for service, and it aims at covering the lack of research on these conversational patterns in telephone talk.

Service encounters on the telephone represent a well-defined category among the whole range of phone calls possible. Both caller and receiver of these calls expect the other to do certain kinds of activities. On the one hand, callers are the initiators of the interaction and, to a certain extent, they are at an advantage with respect to their interlocutor in any phone call: 'The callers know who they are trying to call, and why they are doing so. The answerer, upon picking up the phone, knows nothing on either count' (Hutchby, 2001, p. 111). Callers in service encounters, as well as in all work-related conversations, call for a precise reason, because they need something, and their expectation is to satisfy their request. On the other hand, receivers are usually interrupted in their ongoing activities by the ringing of the telephone (Hopper, 1991). This intrusive character of the phone conversation is sometimes the reason to which speakers refer in the course of the conversation, when they deny or postpone satisfaction of the request.

Receivers at work mostly expect to receive calls related to their activity, either from a 'client' or a colleague, and expect less informal calls from relatives and friends, as 'the distinctiveness of formal types of institutional settings is based on the close relationship between the participant's social roles and the forms of talk in which they engage' (Hutchby and Wooffitt, 1998, p. 149). These should therefore be displayed by the talk produced by the participants in the interaction.

Observing the receivers' responses to the callers' pre-requests for service and requests for information implies looking at them as a result of what was said before, and precisely what the caller is looking for, how s/he formulates the request, either with a closed or an open question,

Introduction 3

and how much information s/he provides to allow receiver (R) to understand what is being asked. We also need to consider how R provides the response: whether s/he constrains herself/himself to what has been requested and provides minimal and simple responses, or whether s/he extends her/his response by providing an account for not having the service requested, or whether s/he provides additional information about the service requested, and so on, as will be described in the following chapters.

The analysis proposed in this book is a contribution to the studies in what has been called 'applied CA' (ten Have, 2001) as it tries to apply findings of 'pure CA' to data in specific conversational settings. Conversation analysts' interest initially was in the pure structure of the conversation, which is inherent in any type of talk, and its social mechanisms. This was the intent of the early studies by Sacks et al. (1974) who started from an institutional kind of data with the aim to explain the organisation of talk-in-interaction as such. 'Applied CA' interest falls into studying distinct types of interaction, such as that in the workplace, applying the concepts discovered by pure CA.

The study of talk in institutional settings has been quite 'common' in the field of conversational studies. Past studies have focused on talk in courtrooms (Atkinson and Drew, 1979), classrooms (McHoul, 1978; Mehan, 1979), medical consultations (Silverman, 1987; Sarangi and Roberts, 1999), interviews (Greatbatch, 1988), service encounters (Aston, 1988a; Zorzi, 1990; Brodine, 1991; Luke and Pavlidou, 2002; Thüne and Leonardi, 2003; Vinkhuyzen and Szymanski, 2005; Lee 2011a, b), and emergency calls (Whalen and Zimmerman, 1987; Zimmerman, 1984; Monzoni and Zorzi, 2003). Such studies have led to the definition of talk in institutional settings as differing 'from ordinary conversation in *systematic ways*' (Hester and Francis, 2001). Drew and Heritage (1992) distinguished some particular features that mark the conversation in institutional settings. They outline three main points:

1. Institutional interaction involves an orientation by at least one of the participants to some of the core goal, task or identity (or set of them) conventionally associated with the institution in question. In short, institutional talk is normally informed by *goal orientations* of a relatively restricted conventional form.
2. Institutional interaction may often involve *special and particular constraints* on what one or both of the participants will treat as allowable contributions to the business at hand.

3. Institutional talk may be associated with *inferential frameworks* and procedures that are particular to specific institutional contexts (Drew and Heritage, 1992, p. 22).

In the data analysed for the present study, speakers orient, on the whole, to a particular kind of task, that of providing a service, and their goal is to win a new customer by offering a good-quality service. Moreover, speakers seem to make reference to a shared framework, or schema, for the way in which the responses should be developed. In addition, this work belongs to 'applied CA' studies as its findings may be used to advise people and organisations about how specific interactional practices might be handled in order to facilitate smooth and effective practice (ten Have, 2001).

The sequence of talk taken into consideration in this volume provides an overview of how speakers in these encounters deal with the business of the call. In fact, this sequence belongs to the body of the conversation, and borders on one side with the opening sequence and on the other with the closing one. The request–response sequence occurs in a position in the conversation that is less constrained by communicative rituals, although it is constrained by the sequential shape of the adjacency pair (request–response). In this book I will aim to respond to the following research questions:

- What is the conversational architecture followed in responding to pre-requests for service and requests for information in service encounters? What are the strategies employed by speakers to accomplish these actions over the telephone?
- Do speakers of the three languages under consideration, namely English, German and Italian, share the same set of strategies in performing this action sequence?
- Does the institutional and formal setting of this kind of conversation lead to a specific way of dealing with the conversation? To what extent are the speakers constrained by their institutional and social roles? Can we therefore talk of a specific conversational genre concerning service encounters?
- Given the fact that responses to the requests in the calls analytically observed were systematically expanded, what is the grammatical configuration of such extensions to the response turns, and what do different types of constituency imply for the interaction and turn-taking system?
- What practical implications can be drawn for the training of service providers and call centre operators from the analysis of the

conversational strategies and sequences employed in naturally occurring service encounters?

The various chapters in the book will address the different aspects these questions imply, which will be then put together and summarised in the final Chapters 7–9. More specifically, Chapter 7 will explore the overall organisation of the responses to the pre-requests for service in terms of types of turns and strategies employed, the cross-cultural comparison, and the overall analysis of the different grammatical constituency of response turn extensions. Chapter 8 looks into call centre training issues starting from the data analysed. Finally, the issue of the constraints produced by the institutionality of the talk is addressed in Section 9.1.

The chapters of analysis are organised around the various formats of response that were found. These are preceded by an introduction to the literature on CA and its main analytical concepts. The turn-taking system, the concepts of adjacency pairs and preference and repair are introduced, as well as studies on interactional linguistics, with the relevant studies on turn extensions and a brief discussion of the various approaches on social interaction; these are contained in Chapter 1. Chapter 2 presents the data used for the analysis and summarises the methodology employed before proceeding with the analysis chapters.

The data analysis is divided into four chapters according to the different formats of response available to speakers of the three languages. Within each chapter the format will be treated firstly as a conversational phenomenon in general, and at a second stage a distinction among the different uses of the format in the three languages will also be made. A final section of each chapter will be dedicated to the observation and summary of the use of the different formats of response in the three languages for a partial cross-cultural comparison, and a summary of the different grammatical configurations of turn extensions is also provided.

The analysis of the different response formats will start from the less structurally complex format, called the 'simple response format', and will gradually increase in complexity and use of dispreferred features. Chapter 3 presents the less structured format, namely responses formulated with a simple structure, i.e. which contain just the minimal response to the request made by the caller. Chapter 4 deals with a more complex format of response, that made of the response to the request followed by more talk, and therefore expanded. Chapter 5 displays an even more complex structure, as here responses are at first put on hold by the initiation of insertion sequences aiming at specifying or asking for confirmation of the object of the request. The conversation then proceeds by producing the response to the initial request that can be displayed both in a simple format and in a more elaborate one, with a further extension to it. Finally Chapter 6 looks at a particular phenomenon displayed in part of the calls: the receiver responds to the request in a simple way, similarly to how s/he does in the calls of Chapter 3, but here the request is not dismissed and participants do not proceed to the closing sequence, but the sequence is kept open by the caller's asking more questions and prompting more talk by the recipient. Chapter 7 then summarises the main results; it sums up the different formats of response used by the speakers of the three languages and the grammatical configurations of the response extensions. Chapter 8 discusses the implications for the training of call centre operators and service providers according to the results of the analysis. It provides concrete examples of good and bad practices of service providers, with suggestions for how these could be improved. Finally, in Chapter 9, service encounters as a specific type of institutional talk are discussed, together with some methodological and practical implications that can be drawn from the analysis of this kind of data.

In the Appendix the complete texts of the transcriptions of the calls discussed in the chapters are provided.

1
Theoretical Framework

The theoretical approach of this book is mainly drawn from conversation analysis (CA) and the more recent explorations of this framework within linguistics, such as the research on grammar in interaction. This chapter reviews the main concepts of this theoretical approach, beginning with the turn-taking system and adjacency pairs. Studies on the grammatical constituency of turns and turn expansions are also reviewed. The description of these concepts is useful to set the conversational features found in the data in their wider context and application to a specific conversational setting, i.e. of service encounters. In this way it will then be possible to describe which features common to all talk-in-interaction and inherent to it also belong to the talk in service encounters, and which are the specific characteristics of these encounters. This chapter examines firstly one of the core concepts in CA, the turn-taking system.

1.1 The turn-taking system in conversation

The turn-taking system is a core concept in the studies on social interaction and naturally occurring speech. Sacks et al. (1974) in their seminal paper proposed a mechanism that gave an account of how talk-in-interaction could take place smoothly without overlaps and long gaps between the turns. Therefore they proposed that the conversation between two or more participants is organised in turns, implying that speakers take turns in speaking in an orderly way. The ongoing conversation is thus established by the sequential exchange of turn-taking between participants, whenever the previous discursive unit is complete. The simplest systematics for the organisation of turn-taking (Sacks et al., 1974) describes the construction of turns of talk into units. Each turn unit has two components: the turn constructional component and the turn-allocation component. The constructional component (TCU) includes syntactic and intonational units: words, phrases, sentences, etc.: 'Instances of the unit-types so usable allow a projection of the unit-type under way, and what, roughly, it will take for an instance of what unit-type to be completed' (Sacks et al., 1974, p. 702).

Conversation analysts, however, do not intend to define what a TCU is a priori. Speakers, in fact, manage and negotiate each unit during the course of the conversation (Gavioli, 1999, p. 45). According to the model, when the current speaker reaches the completion of a TCU (possible completion point, PCP, which therefore implies a transition relevance place, TRP) a speaker change is possible. The allocation of the new turn may take place:

- By the current speaker selecting the next speaker;
- By the self-selection of the second speaker;
- Or, by the current speaker continuing to speak.

Near the TRP, pauses or slowdowns may occur if no speaker is selected or self-selected for the allocation of the next turn. Short overlaps between the turns or simultaneous starts can be found when one of the speakers self-selects for taking the next turn. In the last instance the rule followed for the allocation of the turn is the speaking first rule.

TCUs and TRPs seem highly dependent on one another, creating a relationship of dependence of the TCU upon the TRP. In the last few years such dependence of one component upon the other has become unclear to researchers. Selting (2000) has suggested possible clarifications of this notion and a distinction between the two components. The original conceptualisation of TCUs in fact, relying on TRPs for deciding on their completion, did not give a full account of longer turns in conversation that could be accounted for as being a multi-unit since they meet various prosodic contours and syntactic completions. Selting (2000) therefore suggests, in her analysis of 13 episodes of storytelling, that speakers participating in 'larger projects' align to the existence of several units that reach their completion (TRP) but do not necessarily imply a speaker change. What changes in the system here is that TRPs occur and mark the end of TCUs, but do not necessarily and always imply turn-ending and speaker change. Rather, the preceding unit may be continued by the addition of more talk, syntactically and prosodically integrated with the preceding turn unit or displaying a new prosodic unit:

> If this expansion of the inherently and fundamentally flexible TCU or turn is displayed as prosodically integrated, speakers will present it as the continuation of the prior TCU. If this expansion is displayed as prosodically exposed in a new prosodic unit, the speaker will present it as a new TCU. Every complete turn is by definition also a TCU, but not every TCU is a possible turn. (Selting, 2000, p. 512)

With such a proposal, the system proposed by Sacks et al. (1974) remains the same for turn allocation, relevant at each TRP, and modifies only the turn constructional component. Moreover, Ford and Thompson (1996) have contributed to a better understanding and description of the factors playing in the construction and definition of a TCU, which include:

- Syntactic completion providing projectable units, as the Sacks et al. model already suggested. Ford and Thompson consider as syntactic complete utterances, and therefore possible units, 'a point in the stream of talk "so far", a potential terminal boundary for a recoverable "clause-so-far"' (Ford and Thompson, 1996, p. 143).
- Intonational completion, by which is meant 'a stretch of speech uttered under a single coherent intonation contour' (Du Bois et al., 1993).
- And they add to the Sacks et al. (1974) model a point of pragmatic completion. This means that an utterance needs to have a final intonation contour and has to be interpretable as a complete conversational action within its specific conversational context.

A TCU is therefore considered complete when the three factors, i.e. the syntactic, intonational and pragmatic factors, reach a point of completion. These play an important role in the participants' projection of the upcoming occurrence or non-occurrence of transitional places in advance (Ford and Thompson, 1996, p. 171).

The data analysed in this book display the production of both simple and more complex TCUs that are often expanded or momentarily interrupted by the occurrence of insertion sequences clarifying previous talk. The allocation of turns is made by using all the possible ways of turn allocation. Speakers select one another when they initiate sequences such as requests that leave the floor to the other speaker to respond, as in the example below (lines 03 and 04):

Example I: Florist, Cologne

01	C:	((telephone rings))	
02	R:	blumen meier guten tag?	R: meier flowers good morning?
→ 03	C:	guten tag julia schnibben (.) ehm ich wollte fragen? ob sie mir- ehm etwas über die pflege von bonsaibäumen sagen können.	C: good morning julia schnibben (.) ehm i wanted to ask (you)? if you could tell me- ehm something about the care of bonsai trees
04	R:	bonsai führen wir gar nicht. da gibt's in der innenstadt? sind sie von kö-öln?	R: actually we don't sell bonsai. there is (one) in the city centre? are you from kö-öln?
→ 05	C:	ja	C: yes
06	R:	am stadtmarkt ist ein geschäft [...]	R: there is a shop in the city market [...]

(CVD 4 *K22F *K Blumen2)

The allocation of the turns also takes place by a self-selection of the next speaker, as done in the next example (II) by C in line 07 in order to let the conversation continue: after R has responded to the first of her questions about prices, she then retakes the floor to assess the just delivered news, and then proceeds to ask for further information:

Example II: London, Beauty Centre

01 C: ((telephone rings))
02 R: (an absolute look) how may i help?
03 C: .hh hello i was wondering could you tell me how much it costs fo:r a manicure please
(0.3)
04 R: a manicure?
05 C: mh[m
06 R: [fifteen pounds

07 C: okay >and do< i have to make an appointment
08 R: (we wont make an appointment) but if you come in when i'm busy (then i will fit you in)

(SNGB 04 *IR24F1 *LO BEAUTY CENTRE)

Finally, one speaker can decide to continue talking, as R in example I does in her response turn, line 04, where, after imparting the information in response to the request for information, she does not deal with what is requested; she produces this chunk of talk as interactionally complete from a syntactic, prosodic and pragmatic point of view, and continues talking, by suggesting a place where C could go.

1.1.1 Adjacency pairs and sequence organisation

Participants in the conversation also perform actions in the turn-taking exchange, which are usually mutually accomplished and occur in pairs of turns. For instance, if one speaker produces a question, with such an action s/he presupposes and constrains the other speaker's answer to it. Only when the question has been answered can the action be considered complete. Such actions that are complementary to one another are called adjacency pairs: 'Adjacency pairs are really minimum joint projects. What A is doing in asking B a question is projecting a task for the two of them to complete – the exchange of information specified in her question' (Clark, 1994, p. 992).

There are various activities that are paired in a conversation. Schegloff (1968, 1972) has distinguished some of the possible pairs: summons/answer; question/answer; closings; invitation/acceptance–decline; offer/acceptance–decline; complaint–apology/justification. Adjacency pairs have similar formal characteristics, some of which are listed below (cf. Schegloff and Sacks, 1973, pp. 295–6, and Psathas, 1995, p. 18):

1. They are (at least) two utterances in length.
2. They have (at least) two parts.
3. The first-pair part is produced by one speaker.
4. The second-pair part is produced by another speaker.
5. The component utterances are produced in adjacent position, i.e. they are in immediate next turns.
6. The two parts are relatively ordered in that the first belongs to the class of first-pair parts and precedes the second-pair part, which belongs to the class of second-pair parts respectively.
7. The two are discriminatively related in that the pair type of which the first is a member is relevant to the selection among second-pair parts.

12 *Business and Service Telephone Conversations*

8. The two parts are in a relation of conditional relevance in such a way that the first sets up what may occur as a second-pair part, and the second depends on what has occurred as a first.

Adjacency pairs link one pair part to the other through different degrees of projection. 'The strongest projection in interaction pre-structures a conversational slot uniquely by making òne specific next utterance conditionally relevant' (Auer, 2005, p. 16). The strength of the projection varies according to different types of actions made relevant by the first-pair part. This means that the projection will be stronger when the first-pair part makes relevant just one type of second-pair part. The adjacency pair opened up by an invitation projects at least two options, acceptance or declination, and it is therefore stronger with respect to the second pair in response to compliments, as they allow different types of next actions, such as rejection, acceptance, acknowledgement, counter-compliments, etc. (Auer, 2005). The request–response sequence observed in the following chapters displays the same strength of projection and conditional relevance between the parts of invitation sequences; they project two types of second-pair part, satisfaction or non-satisfaction of the request.

The relation of conditional relevance that binds the two parts of the pair refers not only to the fact that given the first (part), the second is expectable, and 'upon its occurrence it can be seen to be a second item to the first', but also that 'upon its non-occurrence it can be seen to be officially absent, all this provided by the occurrence of the first item' (Schegloff, 1968, p. 364). The absence of the second-pair part needs to be justified and is often the result of a misunderstanding or disagreement. It denies the systematic character of the conversation that predicts the regular recurrence of a phenomenon; this is why when this does not occur, there is something in the conversation that signals it should have occurred (such as a justification, and so on). At other times the second-pair part may be delayed and occurs after the completion of another action that breaks the current one, that is the initiation of an insertion sequence (term used by Schegloff, 1972, and then Levinson, 1983) or a side sequence (term used by Jefferson, 1972). Such sequences are initiated between the first- and second-pair part, putting on hold the current activity when clarification for something in the first-pair part is asked for or something is explained further (Schegloff, 1972; Jefferson, 1972).

The beginning and the closing part of a conversation are the places where it is easier to find the pairs in their basic constitution: 'Adjacency pairs occur in talk wherever "ordering is required"; they are specifically "fitted" to the solutions of problems at initiation and termination of encounters' (Hopper and Doany, 1988, p. 162).

But adjacency pairs also constitute other types of more elaborated sequences that can be found in the course of the conversation. We may find the 'four-part structures' made of an adjacency pair preceded by a 'pre-sequence' or 'prefatory sequence', which has normally two parts too. Or there can be 'extended sequences' in which we may have storytelling, or giving directions during the flow of the interaction.

In sum, adjacency pairs provide an armature around which secondary organisations can form and sequences of talk are constructed:

> These organizations can be schematically represented as expansions that are organized in relation to a 'base' adjacency pair. Most of these expansions address the appropriateness of first actions, management of the prospects that desirable second actions will come to pass, and management of situations in which those second actions depart from the expectations (or desires) of producers of first actions. (Heritage, 2008, p. 306)

The responses to the requests for information and pre-requests for service of the service encounters examined in the following chapters display both minimum and wider projects, as the adjacency pair is often either interrupted by the occurrence of insertion sequences or expanded, thus forming more complex conversational sequences.

1.2 Preference organisation

Preference organisation has been claimed to be relevant for the request/answer adjacency pair (Atkinson and Drew, 1979; Wootton, 1981; Levinson, 1983; Heritage, 1984b; Bilmes, 1988; Mey, 1993; Bonu, 1995; Yule, 1996; Boyle, 2000; Schegloff, 2007; Church, 2009). Requests are involved in preference organisation because, as in invitations, offers or proposals, they can receive at least two different types of responses, for example they may be accepted or rejected (Heritage, 1984b).

As Schegloff (2007) notes, only some types of sequences call for one central type of second-pair part. Greeting exchanges belong to this category: once a speaker produces a greeting, in fact, this invites another greeting to be produced by the interlocutor, and no other second-pair parts are relevant in response to it. The majority of sequences, though, like the ones listed above, provide alternative types of response, which embody different alignments toward the project undertaken in the first-pair part (Schegloff, 2007, p. 58).

The main issue in the organisation of preference has thus to do with the alignment of second-pair parts with respect to the first. Schegloff (1988, p. 453) has described two complementary ways in which the concept of preference is used in CA:

- One approach focuses on the turn design of first parts that are likely to prefer certain seconds (Sacks, 1987; Heritage, 1984b).
- The other approach looks at the structure and the design according to which second parts are constructed (Pomerantz, 1984; Levinson, 1983).

With respect to the first approach, preference can be seen as a mechanism of ranking choice over the alternative second-pair parts available, so that there is at least one preferred alternative response and one dispreferred. In Sacks' words (1987), the first- and second-pair parts are linked together by a 'preference for agreement'. The two types of preference are said to be 'talking together', i.e. when an agreement answer occurs, it usually occurs contiguously to the assessment turn. And it therefore responds not only to a preference for agreement, but also to the preference for contiguity. Conversely, if a disagreement occurs, 'it may well be pushed rather deep into the turn that it occupies' (Sacks, 1987, p. 58), and in this way it neither responds to the 'agreement rule' nor to the conditional relevance implicated by the production of the assessment.

Preferred and dispreferred seconds also refer to the structural organisation of the sequence parts. Pomerantz (1984) gives an extensive description of the different second-pair part formats possible in agreements and disagreements to second assessments. She shows that a prior assessment provides the relevance of one of its next actions over its alternatives. The next action which speakers are oriented to is called 'preferred next action', and in the case of second assessments it is an agreement with the prior assessment, and occurs contiguously to the prior assessment turn. Here follows a list of the main features of the two types of sequences, agreement and disagreement formats in second assessments:

> Agreements have agreement components occupying the entire agreement turns; disagreements are often prefaced.
>
> Agreements are accomplished with stated agreement components; disagreement may be accomplished in a variety of forms, ranging

from unstated to stated disagreements. Frequently disagreements, when stated, are formed as partial agreement/disagreement; they are weak forms of disagreement.

In general, agreements are performed with a minimization gap between prior turn's completion and agreement turn's initiation; disagreement components are often delayed within turn or over a series of turns.

Absences of forthcoming agreements or disagreements by recipients with gaps, requests for clarifications, and the like are interpretable as instances of unstated, or as yet-unstated, disagreement. (Pomerantz, 1984, p. 65)

According to these distinctions, preferred and dispreferred turns would respectively look like the examples below:

Example III: (50) Atkinson and Drew (1979, p. 58)

A: why don't you come up and see me some //times
B: I would like to

Example IV: (51) Atkinson and Drew (1979, p. 58)

A: uh if you'd care to come and visit a little while this morning I'll give you a cup of *co*ffee
B: hehh Well that's awfully sweet of you,
((DELAY)) ((MARKER)) ((APPRECIATION))
I don't think I can make it this morning.
((REFUSAL or DECLINATION))
.hh uhm I'm running an ad in the paper and-and uh I have to stay near the phone. ((ACCOUNT))
(from Levinson, 1983, pp. 333–4)

The distinctive features pointed out by Pomerantz have been explicitly translated in preference terms, so that the features displayed in agreements have become the preferred formats and those of disagreements the dispreferred formats. Preferred second-pair parts may be treated as the default or 'response' of reference and they are usually produced straightforwardly after the first-pair part has been delivered (Schegloff, 2007, p. 66), as in example III above. In contrast, dispreferred second-pair parts can contain mitigations or attenuations, elaborations such as accounts, excuses, disclaimers or hedges and they are not produced contiguously to the first part, but they are rather characterised by:

(a) Delays: (i) by pause before delivery, (ii) by the use of a preface (see (b)), (iii) by displacement over a number of turns via use of repair initiators or insertion sequences.

(b) Prefaces: (i) the use of markers or announcers of dispreferreds like Uh and Well, (ii) the production of token agreements before disagreements, (iii) the use of appreciations if relevant (for offers, invitations, suggestions, advice), (iv) the use of apologies if relevant (for requests, invitations, etc.), (v) the use of qualifiers (e.g. I don't know for sure, but ...), (vi) hesitation in various forms, including self-editing.

(c) Accounts: carefully formulated explanations for why the (dispreferred) act is being done.

(d) Declination component: of a form suited to the nature of the first part of the pair, but characteristically indirect or mitigated. (Levinson, 1983, p. 334)

Example IV displays dispreferred features in the response, in which the refusal only comes after the turn has been constructed with a delay, the use of the marker *well*, and the expression of appreciation, and it is followed by the production of an account to justify the refusal.

The sequence analysed in this book is relevant in terms of preference organisation, since alternative second parts are available to R in order to respond to the request made by C.

1.3 Repair and correction

In the production of a turn, speakers may need to 'correct' errors they make or misunderstandings that may arise from the immediate prior talk. This phenomenon also includes cases in which one tries to recover a term or a word, the reformulation of an utterance, the comments a speaker makes on his speech, though errors are not explicit, etc. For the wide domain of this phenomenon, conversationalists prefer to refer to and use the term 'repair' rather than 'correction' in order to capture all its occurrences (Schegloff et al., 1977).

Schegloff et al. (1977) identified a system that seems to be followed in all these trouble-source occurrences. They suggest that there is a preference order in the choice of the correction, and they first distinguish between 'self-correction' and 'other-correction', i.e. correction by the speaker of what is being corrected vs correction by some 'other' (Schegloff et al., 1977, p. 361). In particular there is:

- Self-initiated self-repair;
- Other-initiated self-repair;
- Self-initiated other-repair;
- Other-initiated other-repair;
- Failure from self-initiation;
- Failure from other-initiation. (Schegloff et al., 1977, pp. 364–5)

Both self- and other-repair are strictly related to one another in their performing the same action, and self-initiated repair is normally preferred to the other-initiated one, so that if the correction is not performed in the same turn of the trouble source, the interlocutor usually gives a second opportunity for a self-initiation by repeating the trouble source instead of correcting it (Schegloff et al., 1977).

Repair instances are also present in the deployment of the responses to the requests in the service encounter. Their occurrence is sometimes due to understanding problems, or at other times they occur in R's response in order to repair a lack of service.

1.4 Response tokens

1.4.1 *Oh*-receipts and discourse markers

Turn beginnings represent a significant and strategic aspect in turn design, as recognised by the relevant literature (Sacks et al., 1974; Schegloff, 1987, 1996; Lerner, 1996; Heritage, 2002). Turn beginnings often project the planned shape and trajectory of the just initiated TCU, providing hearers with resources for anticipating the kind of action that is being deployed as well as with what will take the speaker to complete that action (Schegloff, 1987). Turn beginnings are therefore important for the anticipation and organisation of sequences of actions and for the management of the turn-taking through which those sequences are implemented.

Turn beginnings are also strategic sites because they are a privileged location for the placement of discourse markers that convey some relation between what the speaker has been saying and what s/he is going to say next. So turn components like *well, uh, but, so, oh,* and others are used in this way (Heritage, 2002, p. 197).

Discourse markers play a role in the phone calls analysed. Also called pragmatic markers, they have been defined as particles that:

1. Do not affect the truth conditions of an utterance;
2. Do not add anything to the prepositional content of an utterance;
3. Are related to the speech situation and not to the situation talked about;
4. And have an emotive, expressive function (Jucker and Ziv, 1998, p. 3).

They therefore play a function within discourse and 'they select, and then display, structural relations between utterances, rather than create such relations' (Schiffrin, 1987, p. 321).

One marker that is frequently found in the data under consideration here is *oh*, which has been defined as a 'marker of information management' as it 'pulls from the flow of information in discourse a temporary focus of attention which is the target of self and/or other management' (Schiffrin, 1987, pp. 73–4). It is used to mark the production and reception of information, the replacement and redistribution of information and the receipt of solicited, but unanticipated information, and it is more likely to be used when the provided information was not expected by the speaker.

Heritage (1984a, 1998, 2002) has widely explored the use of the particle *oh* in its occurrence in response to some kind of talk as displaying a change of state (1984a). It will be reported here on its use after some kind of information, which is also the case of the data that will be analysed in the following chapters. The production of such tokens after the delivery of information serves to mark the receipt of the informing delivered in the preceding talk. Heritage (1984a, p. 301) reports it occurs in response to complete chunks of information and is therefore produced at points at which the information is possibly complete.

Oh is usually found in turn-initial position, and is claimed to be mostly co-occurring with additional turn components, such as assessments or requests for further information (Heritage, 1984a):

> They [*ohs*] are commonly used to receipt answers-to-questions as informative, while withholdings or substitutions of 'oh' receipts may be used to imply either that an answer was not, or not yet, informative or, alternatively, that a prior question-formed utterance did not request information. (Heritage, 1984a, p. 312)

Oh at turn beginnings has therefore been studied as being used to convey a stance toward what a previous speaker has said (Heritage, 2002, p. 197).

When used to preface the response to an inquiry, *ohs* can indicate that the question to which it responds is inapposite in some way, and it can do so inexplicitly and self-attentively. Heritage (1998, p. 296) defines some basic characteristics of *oh*-prefaced responses to inquiry as a conversational practice in saying that they can:

- indicate that the inquiry being responded to is problematic as to its relevance, presuppositions or context;
- foreshadow reluctance to advance the conversational topic invoked by the inquiry;
- be a component of 'trouble-premonitory' responses (Jefferson, 1980) to various types of 'how are you' inquiries in conversational openings and elsewhere (Heritage, 1998, p. 296).

In contrast, *ohs* are less often found as free-standing in the turn (Heritage, 1984a). In this position, they display the speaker's preparedness to treat the prior answer as complete.

The cross-cultural data considered for this study display the use of the token *oh* and its equivalents in the other two languages under examination. The tokens that seem to perform the same change-of-state function as the English *oh* are *ah* in Italian and *ach* or *ah* in German. Some occurrences of those tokens will be analysed in the following chapters.

1.4.2 *Okays, yeah, mm* and *mm hms*

The above tokens have been studied as uttered by one participant in the conversation in receipt to the talk of his interlocutor. Their function has been studied as being different from one token to the other. Jefferson (1985) has looked into the use of *yeah* and *mm hm*, and more recently Drummond and Hopper (1993) have re-examined the use of both tokens. In contrast Gardner (1997) has researched the use of another token, *mm*, in its different uses and functions. Finally, *okays* were studied by Beach (1993) and Guthrie (1997).

When looking at *yeah* and *mm hm*, Jefferson has described them as both pertaining to the nature of acknowledgement tokens, but has distinguished *yeah* from *mm hm* in that the first is recognised to be used by the recipient to move into speakership, and is therefore defined as a token of speakership incipiency. *Mm hm*, on the other hand, seems to exhibit 'passive recipiency' (Jefferson, 1985, p. 206). Drummond and Hopper (1993) have found that *mm hm* is only used to take the turn but not the floor, which means that such a token is not followed by further talk from the same speaker, and the same holds true for the token *uh huh*. Both *mm hm* and *uh huh* therefore occur as free-standing objects in the ongoing talk. Confirming Jefferson's results, Drummond and Hopper provide further accounts in the description of such tokens as one holding passive recipiency and the other speakership incipiency. Also the token *yeah* has been found to occur followed by other talk from the recipient up until then, and therefore displaying features for turn- and floor-taking.

Therefore, given the above functions and sequential occurrences, as Schegloff (1982) observed, *mm hms* occur in the course of an extended turn during which a recipient shows her/his understanding of such a unit and they show that it is not yet complete.

Gardner (1997) found the use of *mm* by speakers with different prosodic contours associated with different functions, the more common ones being those with falling contour and defined as being a weak acknowledging token. With such a prosodic contour it displays speakership incipiency between *mm hm* and *uh huh* on the one hand, and *yeah* on the other. And similarly to *yeah*, this type of *mm* is then usually followed by the same speaker talk:

> Mm is insufficient on its own to do the work of heralding a topic change. What, however, it does do as a third position receipt token is to convey that its speaker has nothing more to add to the topic so far, that is, it is retrospective and closure relevant, at the same time paving the way for the introduction of a new topic. (Gardner, 1997, p. 135)

The other two main functions of *mm* reported by Gardner (1997) are its occurrence as continuer and as assessment. When it occurs as continuer, it is produced with a rising terminal pitch. In contrast to the previous function, this kind of token is articulated unclearly and perceived as difficult to understand by the recipient. It is relatively neutral in terms of emotional, evaluative, surprised or remarked content and often occurs at a place where the TCU is not complete (Gardner, 1997, p. 143). Such tokens thus orient to partial turn completion, in a similar way as *mm hm* does.

Finally, *mm* displaying assessment is found with rise–falling contour, as other assessments, and among the different types of *mm*, this is the one that displays the highest speakership incipiency, due to a greater involvement:

> With 'mm', the rising part of the contour shows heightened involvement (see Selting, 1994), although the minimalness of the token

makes it a relatively weak assessment. [...] the terminal fall appears to indicate finality, a sense of completion, and receipt of the prior unit without any problems, a characteristic it shares with the falling, acknowledging 'mm'. (Gardner, 1997, p. 147)

The last token to be discussed is *okay* which has been found to occur both in turn-initial position and thus followed by other talk of the utterer, or as a free-standing token. Its occurrence has been identified by Schegloff and Sacks (1973) as a device initiating movement toward closure and/or passing turns en route to terminating phone calls. Beach (1993) has observed *okay* usages as having a dual nature, being both closure relevant and continuative at the same time in that they orient to the closing of the current action and at the same time pave the way for a new action/argument to take place. Instances of such occurrences will be observed in the analysis of the data in the next chapters.

In contrast, free-standing *okays* are found at transition relevance places (TRPs) and at points in conversation perceived as meaningful (Beach, 1993). In response to questions they display affirmative responses. They are also found as third-turn receipts following clarifications, invitations, offers, information giving, etc., thus working as acknowledgement tokens too (Beach, 1993; Guthrie, 1997). Finally, *okay* differs from *mm hm* in that it occurs after an utterance 'which is in some way more complete than the utterances which an "mmhmm" follows' (Guthrie, 1997, p. 398).

The next section will look at another CA-related theoretical approach that has been used in the analysis of the data, interactional linguistics and the grammatical configurations of turns.

1.5 Interactional linguistics

Interactional linguistics is a multidisciplinary research field where scholars from different backgrounds interact. It puts together linguists, conversation analysts and anthropologists. Scholars in this field look at all aspects of language structure and use, i.e. phonetics, phonology, morphology, syntax, lexis, semantics, pragmatics, but also language variation, language acquisition, loss and disorders, and they study them from an interactional point of view (Selting and Couper-Kuhlen, 2001).

A central point in the literature in this field is represented by the aim of understanding each aspect of language itself, the activity itself, and how each of these interacts with its context of activity. Language data are approached not to study each phenomenon by itself, i.e. syntax for the sake of describing a theory of syntax, but to outline a theory of syntax that arises from the data. Studies in this field thus

> vary in the degree to which they explore the mutual bearing of interaction and grammar conceived in a more or less familiar form, on the one hand, and, on the other, the ways in which the intersection of interaction and grammar prompts a rethinking of what sort of thing grammar might be thought to be and how it may be configured. (Schegloff et al., 1996, pp. 27–8)

Grammar is not only a resource for interaction and not only an outcome of interaction; it is part of the essence of interaction itself. Or to put it another way, it is inherently interactional (Schegloff et al., 1996, p. 38).

Research in this field is continually increasing, beginning with the edited volumes *Grammar in Interaction* (Ochs et al., 1996), *Prosody in Conversation* (Couper-Kuhlen and Selting, 1996), *Interaction-Based Studies of Language* (Ford and Wagner, 1996), *Studies in Interactional Linguistics* (Selting and Couper-Kuhlen, 2001), *The Language of Turn and Sequence* (Ford et al., 2002a), as well as single-paper publications (Ono and Thompson, 1995; Ford and Thompson, 1996; Selting, 1996; Ford et al., 1996, 2002b; Auer, 2005; Lindström, 2006; Fox, 2007; Fox and Thompson, 2010, etc.).

Publications in this field may take different approaches of analysis which mainly belong to two categories: firstly, there are studies that start from the language structure and then investigate its deployment in interaction; secondly, one may start from the interactional order and identify linguistic practices systematically associated with it (Selting and Couper-Kuhlen, 2001, p. 9). The present study relates to the works in the field of interactional linguistics as it also looks at turn construction from a grammatical point of view, taking the second of the approaches described. The analysis is directed in particular to the grammatical configurations of turn extensions.

1.5.1 Grammatical configurations of turn increments

Grammar has been claimed to play a central role in the projection of turn constructional units and the distinction of multi-unit turns (Schegloff et al., 1996). This study focuses on those response turns that are extended after a point of possible completion. Goodwin (1979, 1981) has contributed to the study of such increments of turns and has suggested the influence on such a deployment of turns and turn extension of both gaze and syntax, pointing out that these two phenomena involve at least a combination of gaze shift and syntax. As gaze is not involved in the talk-in-interaction that takes place on the phone, the study mainly referred to here will be Ford et al.'s (2002a). In their paper they look at the constituency and grammar of turn increments and distinguish different types of turn expansions. The first one is taken from Schegloff (1996) and called 'extension', or 'added segment' by Goodwin (1981). By this term they both mean 'increments that are interpretable as continuations of the immediately prior possibly completed turn' (Ford et al., 2002a, p. 16). These are semantically and syntactically coherent with the preceding talk, and they can be prepositional phrases, for instance, occurring after a PCP, as in the excerpt below:

Example V: (1)

Bill said that he was at le̲a̲st goin' e̲i̲ghty miles an ho̲u̲r.
→ with the t̲w̲o̲ of 'em on it.

(from Ford et al., 2002a, p. 16)

In such instances as the above, the current speaker reaches a point of possible grammatical, prosodic and pragmatic completion, but continues speaking after this point, syntactically attaching his further talk to the previous. Such expansions are thus defined constituents of prior turn units and clauses and continue the action initiated with the preceding talk. They pursue an uptake by continuing the action of the just possibly completed turn, and they thus also provide a solution to a lack of displayed recipiency.

The second type of turn expansion described by Ford et al. (2002a) consists of further talk after a point of possible completion that is not syntactically attached to the previous stretch of talk. These increments are therefore not constituents of prior turn units, and being independent from it, they are called 'free constituents'. Ford and her colleagues focus on just one type of such free constituents, the one represented by unattached noun phrases ('unattached NPs'). In contrast, unattached NPs, in contrast to extensions, 'do the functionally separate action of assessing or commenting on the prior turn material', as the speaker does in the following excerpt (Ford et al., 2002a, p. 18):

Example VI: (10)

Curt: That guy was (dreaming).

→   Fifteen thousand dollars [for an original co:rd
Gary: [figured he'd impressed him
                                (from Ford et al., 2002a, p. 26)

The sequential deployment is similar to that of example V, but here, after reaching a point of possible completion and not getting any uptake on the recipient's part, Curt expands his turn through an NP that further increments the previous talk. This time the increment is not completing the action performed in the previous TCU, but adds to it with a next action, commenting on what has previously been expressed.

The two increment types described by Ford et al. (2002a) thus both have similar characteristics and different ones. Both extensions and free constituents:

- Occur in the environment of a lack of uptake at a TRP;
- And provide a second TRP at which the recipient could display recipiency (Ford et al., 2002a, p. 25).

But they differ in performing different actions:

- Extensions continue the action of the extended turn, often by further specifying when, where, or with whom the related event took place (Ford et al., 2002a);
- Whereas unattached NPs display an assessment and stance with respect to the referent, they offer a standard toward which the recipient could orient in producing the response, a display of the sort of response the speaker is pursuing (Ford et al., 2002a).

This book will analyse response turn increments and look at both extensions and free constituents, and not just at unattached NPs, but at the various types of independent increments.

1.6   Approaches to conversation

CA is obviously not the only way of approaching talk-in-interaction in linguistics. The choice of this theoretical and methodological approach lies in particular in the interest in understanding in a more comprehensive way how we normally organise our speech, and, in Sacks et al.'s words (1974, p. 700), 'trying to understand how conversation can happen at all'.

The closest approaches to CA may be said to be ethnography, interactional sociolinguistics and discourse analysis (DA). All of them carry similar and different orientations, both methodologically and theoretically. The main differences among these disciplines will be briefly summarised in the following pages.

When looking into the social context of linguistic interaction with the aim of discovering the rules of appropriateness of the speech event, the ethnographic approach focuses on its 'surface', i.e. the social context in which it takes place. When trying to account for 'who says what to whom, when, where, why and how' (Hymes, 1972, in Eggins and Slade, 1997, p. 33) 'the SPEAKING grid provides a necessary reminder of the contextual dimensions operating in any casual conversation' (Eggins and Slade, 1997, p. 34).

CA's aim is different as it aims at describing the actual use of interactional strategies in determinate speech events by giving an account of the architecture of the encounters, and how speakers use it, but has no interest in deciding whether or not such interactional strategies are appropriate for the event at hand. CA takes an emic perspective, i.e. 'the perspective from within the sequential environment in which the social actions were performed, and not merely the participants' perspective' (Seedhouse, 2005, p. 252). Detailed ethnographic characterisations are not the primary aim of CA studies, as they are not considered essential for the analytic description of the event and its organisation.

Interactional sociolinguistics also contributes to enriching conversational studies by giving importance to context in the production and interpretation of discourse. In its detailed analysis of the grammatical and prosodic features of interracial and interethnic interactions (Gumperz, 1982) it is useful and relevant for the cross-cultural approach of the present study, such as in the analysis of the use of discourse markers in the three languages.

CA's approach to context is very narrow, since it consists of the immediately preceding talk available to the interactants at the moment of talking. Context is primarily what is relevant to the participants in their target event, 'and not what is relevant in the first instance to its academic analysts by virtue of the set of analytic and theoretical commitments which they bring to their work' (Schegloff, 1992, p. 196). It is said that

> each action is context-shaped in the ways in which it is designed and understood by reference to the environment of actions in which it participates. And it is context-renewing in the way that each action, in forming a new context to which the next will respond, will

inevitably contribute to the environing sequence of actions within which the next will be formed and understood. (Heritage, 1984b, p. 280)

Contextual features may be invoked in the analysis to inform it with the information relevant to the participants at that moment in the interaction and that can explain the talk they are making. The analyst's interpretations 'aim at understanding human social behaviour in its own terms, favouring an emic view close to that of the participants' (Pallotti, 2007, p. 54).

What CA is not interested in is the analysis of talk in terms of speech acts as carrying a particular illocutionary force, according to which each utterance is the realisation of the speaker's intent to achieve a particular purpose that in many instances may be directly derived from the linguistic form of the utterance itself:

> Speech act theory, by treating intention and context as given a priori, treats the utterance as a unit which is recited as a monolithic whole. This view has also been challenged by ethno-methodologically-based work which argues that the utterance itself is largely a joint production, whose form is interactively determined through continuous feedback (Goodwin, 1981). Conceived as a collaborative strategic activity, the discourse process thus becomes one in which both participants are continuously active in negotiating satisfactory outcomes, and the product reflects this continuous collaboration below as well as above sentence level. (Aston, 1988a, p. 3)

CA does not look much at the speaker's intent while performing a turn, but rather at what the speaker is doing in that particular moment and how he is doing it. At the same time, CA differs from systemic functional linguistics (SFL), which is based on the notion of language function and accounts for the syntactic structure of language, placing the function of language as central (what language does, and how it does it). The CA approach wants to observe the structure of conversation as it is seen by participants, accomplished through turn exchange, and how language is used first of all 'here and now'. Only after such analysis can there be an application of the local use to the general structure of conversation. In this sense CA seems to go beyond what SFL does, by giving a broader analysis of the event taken into consideration.

What CA and DA have in common is that they both focus on language as social action. But on the other hand, they are 'divided' by different orientations. DA's action orientation is broader than CA's. The former looks at the wider social functions served by a passage of talk and at the linguistic repertoire provided (metaphors, figures of speech, etc.), whereas the latter's attention focuses on the local management of turns (Wooffitt, 2005). Moreover, the two disciplines focus on different kinds of events: DA gives more attention to disputes or controversial events, such as the academic debates among researchers about competing merits (Gilbert and Mulkay, 1984), whereas CA tends to focus more on the management of routine activities, as has been shown in the earlier pages, turn-taking, sequential organisation, repair, etc. (Wooffitt, 2005). A further point of divergence between the two disciplines is the source data employed for the analysis: CA looks at the data, audio or video, and treats the transcription only as a practical and handy aid; DA primarily looks at texts.

1.6.1 Critiques and debates

The differences among these language-oriented disciplines have inevitably led to academic debates and critiques. CA has its weak points, which have been analysed in depth in linguistics. Discussing them in detail would go beyond the scope of this book. Nevertheless, some of the main points will be assessed.

An interesting debate between Billig and Schegloff was published in *Discourse and Society* (1999). The main issue that embeds all points of the discussion is the different approaches of CA and DA, and a critique of the former for claiming not to theorise a priori the encounters that will be analysed. In fact, Billig (1999, p. 573) criticises CA's claim of being based on a naïve epistemology and methodology. This includes various aspects of the analysis, such as the treatment of speakers' identities. The use of prior categories is acknowledged by Schegloff too (1999, p. 566). The categorisation of the speakers as callers and receivers, for instance, does not deprive them of their personal identities, but deals with them in their sociological position at the moment of the analysis. Further personal details about the participants are not necessarily ignored, but are not relevant and do not guide the analyst's observation in CA.

Another issue of debate is the method of analysis. CA looks at the very details of the interaction, often by looking at single cases or small collections of data that do not aim at larger generalisations. Some researchers consider this kind of analysis as almost useless because they cannot even give an account of what the conversation in a particular field looks like. Conversation analysts defend such analytical procedure by taking a different perspective and saying single case analysis allows the analyst to locate and describe participants' orientation to the events they produce and encounter in their interaction – displays that show how they understand a prior action and how they assign import and consequentiality to it. (Zimmerman, 1993, p. 189)

A third critique made by researchers from other disciplines but also by conversation analysts themselves is that some of the conversational concepts found by CA pioneers are sometimes not entirely operationalisable, and their definitions remain fuzzy. This has been the case of the above-mentioned TCU notion, which has been further explored and clarified by various studies (Schegloff, 1996; Ford and Thompson, 1996; Selting, 2000).

1.6.2 Relevance to this study

All the approaches briefly outlined above share with CA an interest in social interaction and language. This study has chosen to use CA, as the main interest was to provide a description of the interactional organisation of service encounters on the phone. CA's attention to the interactional architecture of the encounters seemed the most appropriate approach for such an aim.

Ethnography of speaking, though looking at the broader context of the encounters, does not fully respond to the needs of performing the kind of analysis necessary for our purposes. Moreover, this book does not aim at assessing the appropriateness of the use of certain interactional strategies in providing responses to the request for information, but at describing all the possible strategies that can be used to respond in this particular conversational setting and in three European languages. Ethnographic details about the participants have been collected and used together with the primary data (the recordings) when they were relevant for the explanation of phenomena produced by the participants in the interaction. Interactional sociolinguistics has provided a contribution to the study of some phenomena, such as the use of discourse markers.

2
Data and Methodology

This book aims primarily at analysing the request–response sequences in telephone service encounters made in three different European countries. It aims in particular to observe the strategies employed by people receiving the calls when responding to the caller's (C's) requests for information. A second aim of the study is that of observing the use of those strategies of response-turn design from a cross-cultural perspective and analysing whether they are shared by the speakers of the three languages. Thirdly, the study aims to look to what extent the participants in the interaction are constrained by the institutional and social roles they play in the conversation, and whether those interactions can constitute a specific conversational genre. Fourthly, the study also aims at analysing the grammatical constituency of extended response turns from an interactional linguistics point of view, and it looks at the implications of such a constituency on the turn-taking system. Lastly, the analysis conducted also aims at providing material from naturally occurring speech that informs the practices of service providers and call centre operators with practical implications for the training of these staff.

When one talks about conversational strategies and the way in which the participants mutually co-construct the interaction, conversation analysis (CA) allows the analyst to look deeply into the interaction and the local organisation of the talk being produced by the participants in terms of turns and sequences. Such analysis shows the strategies used by the receivers (Rs) in responding to C's requests in a detailed way, by faithfully reacting to the participants' moment-by-moment interpretation of the actions carried out through talking. CA methodology is well suited to respond to the other research questions that are addressed in this book. CA's micro-analytic approach is well suited to inquire about the specificities of the conversations under consideration as it belongs to a specific conversational genre in contrast to ordinary conversations. Finally, CA is also a good starting point for tracing an analysis of the grammatical constituency of the response turns. As stated in Chapter 1 (Section 1.5), studies in interactional linguistics attract researchers with different research perspectives to study the interfaces of each aspect of language together with its context of activity. Among the researchers working in this field, conversation analysts contribute to the analysis of language, observing the design of conversational turns and looking at the linguistic practices that can be associated with the construction of such turns. In particular, this book will look at how the construction of response turns is associated with specific syntactic constructions, and how these can influence the organisation of the response itself.

In the next sections of this chapter the data used for the analysis of the aspects listed above will be introduced, and the analytical methods and criteria used in the analysis will also be specified.

2.1 The data

The data considered for this study are a particular kind of institutional talk, one that occurs between customers and/or users on the one hand, and staff members and clerks on the other. Calls were made to small businesses such as shops, offices, travel agents, etc., with the prior consent of the participants, and dealt with a request for information (relevant to the place called).

The data analysed in the present study are taken from a larger corpus collected within a project run at the University of Sassari (Italy) between 2001 and 2006, which includes calls in different languages, such as English, German, Italian, Spanish, French and Russian. The project aimed at analysing telephone calls from both a cross- and intercultural perspective. (The project coordinators were Gabriele Pallotti and myself, and a number of final-year students contributed to the data collection, transcription and analysis of opening sequences in the different languages mentioned above.)

The analysis proposed in this book is based on a random selection of the calls collected within this project. The study is based on a corpus of 141 calls: 50 in British English, 42 in German and 49 in Italian, directed to various services as listed in Table 2.1.

All calls were made between native speakers of the language. All service providers involved in the project were small to medium size, and they employed two to four assistants or clerks. All these businesses have

Table 2.1 Services called in the data

Services called (N = 141)	British English (N = 50)	German (N = 42)	Italian (N = 49)
Hairdresser's and beauty centres	5	4	8
Bookshops, stationer's	13	4	8
Travel agent	4	3	7
Shops (selling clothes, fishing gear, wine, antiques, greengrocer's, car dealers, bars and restaurants)	9	19	12
Chemist's – optician	7	1	8
Medical (doctor's, dentist's, vet's surgery)	3	2	3
Museum – school	3	3	2
Estate agent – bank	2	1	1
Florist	4	5	0

to do with the public both in presence and on the phone. Obviously, some of the services taken into consideration are more used to dealing with customers over the phone than others. The contextual, institutional features specific to each type of service in each of the three countries are relevant for the subsequent format of each encounter. In particular, among the services taken into consideration in this study, these features are:

- Travel agencies in the three countries find it very common to deal with customers over the phone.
- Bookshops also often work over the telephone, although only in Great Britain is it possible to actually order a book, whereas in Italy and Germany the customer would have to go to the shop to do so, as a deposit is commonly required.
- Doctors, dentists and vets are the other category of services that also commonly deal with patients over the phone, most of the time receiving calls asking for times of consultancy or for making appointments, and as for travel agencies there are no relevant cross-cultural differences to note.
- Hairdressers also receive calls quite often in the three countries, although there are some differences when it comes to making an appointment over the phone before getting the service proper. A difference in the kinds of requests that can be made to hairdressers in the three languages is that, whereas both in Germany and Great Britain they usually have apprentices working and practising at

discounted prices in the same hairdressing salon, in Italy this does not happen, as there are specific schools for apprentice hairdressers that also promote discounted prices for haircuts and treatments. So calls in Italy are more likely to ask about discounts and fidelity cards instead, besides asking for an appointment.
- Florists mostly receive calls in which they deal with pre-requests for service, that is information prior to purchase and requests for advice on plant and flower care as well. Florists recorded in both Germany and Great Britain also offer the Interflora service that delivers flowers to homes in different cities in the two countries or abroad. Normally this type of service also requires the customer to go to the flower shop and actually order the flowers and pay for them, unless the internet is used. None of the calls recorded and directed to florists actually requested an Interflora delivery over the phone, but rather only information on the service and its prices.
- Museums and schools are used to dealing with the public on the phone, mainly for giving service information, such as opening hours, facilities or special offers, the types of requests that they actually received in the calls collected.
- The other shops of various kinds, clothes, chemist's, car accessories, antiques and estate agents, are less used to dealing with customers on the phone and they might receive a wide range of requests for information from opening hours to specific items.

Contexts such as those listed above are therefore more likely to receive pre-requests or requests for information that are actually inquiring about a specific service, but do not directly deal with the goods/service exchange. Most of the calls will be followed up by the actual goods exchange at the shop, with a face-to-face encounter. In our analysis the term 'request' can mean request in general, request for some information or pre-requests for a service. The corpus taken into consideration does not aim to be representative for each language, but all results and attempts at generalisation that will be made in the following chapters are indicative trends that can be reconfirmed by analysis on larger samples of data.

2.1.1 Data collection
Data collection followed the scientific method in order to ensure respect for the ethical rights of the participants on the one hand, and naturalness of the data to be collected on the other. Both Rs and Cs gave their assent to be recorded and the data to be used for research

purposes before the data collection took place. Rs were the staff in the various services involved, and Cs were the researchers themselves and their acquaintances. The audio recorder was placed at the C's end. In particular the calls collected in Italian were made by the researchers in the group including the author and her friends and relatives, all native speakers of Italian. In contrast, the calls collected in Germany and Great Britain were made only by acquaintances of the researchers, native speakers of each language. When the calls were responded to by non-native speakers they were not included in this study. Cs did not make up their reasons for contacting a shop or surgery, but only called when they actually needed the service. For this reason the Rs involved were mostly services the Cs had visited before; they did not have any relationship with the shop assistants.

The procedures followed ensured, as far as possible, the natural context of the interactions collected. Rs gave their assent to our team to record some of the incoming calls in a period of time lasting between two weeks and three months, but were not told precisely which calls would be recorded. If Rs knew specifically which calls were going to be recorded, they would pay more attention to the way they were answering the phone and the calls could lose their naturalness. Despite all limitations on spontaneity of the participants, above all on the part of Cs, that could be more biased by the presence of the recorder next to the phone, though, the data collected showed that participants treated the encounters as natural. The requests made by the Cs were therefore perceived as spontaneous, with a few exceptions in which both participants were aware of the unusual request and they made it explicit in the conversation. One team member was responsible for the data collection, and s/he was also the one who facilitated the recording process, by asking for permission for the recordings from all the participants. They also made some of the calls if in Italian.

The calls in English were all recorded in the UK, in Portsmouth, Lancaster, Manchester and London. The German calls were collected in Cologne and Frankfurt and the Italian ones in Sardinia (mainly around Sassari and Nuoro) and Emilia Romagna (Ferrara and Bologna). All Rs' identities were changed into pseudonyms; the same was done with the Cs' identities, when they provided one. Data were transcribed using the Jefferson transcription system provided at the beginning of this book (cf. Atkinson and Heritage, 1984; Hutchby and Wooffitt, 1998; ten Have, 1999; Schegloff's online tutorial on transcription; Charles Antaki online introduction to transcription).

2.2 Research methodology

2.2.1 Qualitative methods

The focus of this study is on R's responses to the requests. As reported by Vinkhuyzen and Szymanski (2005), the response to a customer's request in business is a delicate matter, above all when the response does not satisfy the request. 'Economic analysis has shown that the negative consequences of dissatisfying a customer can be much costlier than the positive consequences of satisfying a customer' (Hart et al., 1990 reported in Vinkhuyzen and Szymanski, 2005, p. 91). Lee (2011a, p. 111) also stresses such possible negative consequences, which may also lead to additional costs for the organisation at hand.

The response to the requests is therefore a delicate matter for the shop assistant to provide, as through it the person calling may decide to go or not to go to the service called and buy or use the service they offer. In contrast to Vinkhuyzen and Szymanski (2005) and Lee (2011a), the attention of this study was devoted to all responses, both satisfying and non-satisfying. In line with other CA work, the methodology employed in the first instance for the analysis of the data had a qualitative basis. Data were approached without any preconceived idea (ten Have, 1999) about the 'what and why' of speakers' actions. Each conversation and, more precisely, each conversational sequence containing the request and its response were analysed. The analysis procedure consisted of four steps:

- At first, each phone call was examined, and the excerpt identified, which consists in identifying the borders of the request–response sequence, from its beginning to its end.
- Then the turn-taking system of the sequence was analysed, and turn construction, pauses in and between turns, overlaps, etc. were identified.
- Thirdly, the analysis focused on the construction of the sequence that is the production of both adjacency pair parts in the request–response sequence. More attention was dedicated to the second-pair part, which is R's response, and the way it was formulated and different response formats were identified.

All the concepts of analysis employed (turn-taking, etc.) have been described in more detail in Chapter 1, as they represent both theory and methods of analysis in CA. As in all conversational work, the question that was being answered through the whole process of analysis was always that of giving an account of what participants were doing moment by moment.

2.2.2 Criteria for the classification of turn increments

Once the request–response sequences had been identified and the analysis focused more deeply on the second-pair part, a distinction was made between the actual and relevant response to the request made by C and the subsequent talk, i.e. an expansion of the turn unit just produced. The extension of the response has an impact on the overall conversational exchange and on the shape of such encounters. An aspect that contributed to the understanding of the function of such expansions in the response turns was the observation of their grammatical constituency with respect to the previous stretch of talk. The analysis focused on this specific aspect of response expansions following the classification of turn increments outlined by Ford et al. (2002a) and described in Chapter 1. As previously mentioned, their work focused primarily on the observation of one kind of turn increments: unattached noun phrases (unattached NP, Ford et al., 2002a). The interest of that study was that of looking at all types of turn increments. A distinction between increments syntactically independent from the previous stretch of talk and 'proper extensions' (as Ford et al., 2002a name them from Schegloff, 1996) was made. In the latter, they talk about turn expansions that continue the preceding talk, mostly clauses attached to the response. Increments were identified as soon as the relevant response to the request was delivered and further talk of R followed:

- After a short pause with no uptake of the caller;
- Or after a pause, which in turn was followed by a receipt token by the interlocutor;
- And by a change of intonation in the receiver's talk.

Free-standing increments were defined as all clauses constructed independently from the initial response, including paratactic constructions. Extensions were defined as all dependent clauses, i.e. subordinates. These two types of response expansion were subdivided according to their occurrence either in the same turn of the response or in another. When it is in another turn, the expansion often comes after a receipt token by the interlocutor such as *mm hm, yeah, okay*, or a repetition of the delivered information. We therefore have four classes of turn increments:

- Free-standing increments occurring in the same turn as the response;
- Free-standing increments occurring in a different turn from the response;

- Extension increments occurring in the same turn of the response;
- Extension increments occurring in a turn following the one with the response.

The various formats of response, together with the different configurations of turn increments, were observed both from a qualitative perspective and a quantitative one, at a further stage of analysis.

Reference to turn expansions in general will be made by using the term *turn expansion* or *turn increment*. When reference is made to one or the other type of expansion, the terms *free*, or *free-standing increment* or *constituent* will refer to those turn continuations syntactically independent from the core response, and *extensions* will be used to refer to grammatically bound extensions.

2.2.3 Quantitative methods

The use of quantification has always raised discussion among conversation analysts, although it has been acknowledged as representing a new direction of research in conversation analytic projects (Heritage, 1999). Schegloff (1993) has pointed out that the use of quantification cannot be meant to be a substitute for analysis. He stresses the fact that quantification is not a way of avoiding looking closely at the phenomenon, at its moment-by-moment deployment, the organisation and understanding of the participants in the interaction. He also points out another issue: the amount of knowledge we have about a certain phenomenon seems to be a prerequisite for quantification.

Quantification in CA was recently re-evaluated in studies looking at larger and more representative corpora that provided statistical evidence of the occurrence of certain phenomena in naturally occurring speech (Drew, 2005). When working with larger corpora instead of single cases, quantification can be used as a fine-tuning device for grounding and extending minimal or bigger generalisations on the weight of that phenomenon in the conversational exchange. As also Barnes and Armstrong (2010, p. 59) state, 'the application of quantitative analyses to conversation can be problematic if they are not combined with sufficiently rigorous qualitative analyses' (Lesser, 2003). The quantitative approach in the present study constitutes the last step of the analysis. Quantification was used in evaluating the use of the various response formats and the use of different types of turn increments.

Quantification involved a frequency count. All tables from Chapters 3 to 6 display partial values, and show the proportion of data displaying each format of response with respect to the whole corpus. All partial calculations are put together in the tables in Chapter 7, showing the overall use of all formats of response analysed. Finally, the three languages under consideration were kept separate in quantification, so that variation among the different uses could also be observed.

2.3 Conclusion

The analytic approach taken for this study is certainly not uncommon within the CA framework, both in the choice of this kind of data and in the analysis methodology employed.

CA is primarily interested in studying naturalistic data, i.e. in data that are not elicited with experimental procedures and that are sampled in conversations that would take place independently from the analyst's recording (Levinson, 1983; Hutchby and Wooffitt, 1998; ten Have, 1999; Mondada, 2005; Pallotti, 2007). Although both types of data are actually taken into consideration by conversation analysts, the first type of sampling is certainly preferred, since those kinds of data can permit the researcher 'to develop an emic, holistic perspective and to portray how the interactants perform their social actions through talk by reference to the same interactional organizations which the interactants are using' (Seedhouse, 2005, p. 257). The data collected for the present study, although constrained by the ethical procedures described above, represent a sample of what people normally do while calling a service provider and requesting information about the services offered. They are a naturalistic kind of data, suitable for an analysis in terms of conversational practices, since they are the expression of human practices in a specific field, i.e. that of service-encounter conversations.

The use of both qualitative and quantitative methods, including cases when just one of those methods is used, does have some constraints as well as some advantages. In particular the quantitative methods need to be used cautiously, and as stated above, any kind of generalisations that will be made should not be considered as general categories accounting for the whole range of request–response sequences possible in service encounters. All generalisations will be made with respect to the data observed, and further research applying the same criteria of analysis may confirm and strengthen the claims made in this study.

Although 'quantifying implies the identification of clear, neatly defined categories, which are problematic in the study of a complex phenomenon such as talk-in-interaction' (Pallotti, 2007, p. 59), the use of such methods can lead towards more valid studies in this field. The explicit categorisation of certain recurrent phenomena and their subsequent counting may be useful for discovering patterns that might otherwise not be seen if just trusting the researcher's 'intuitive grasp' (Schegloff, 1992) or the observation of deviant cases as opposed to the norm, as is often done in analytic studies of conversation. On the one hand, combining qualitative analysis together with quantitative methods can ensure the internal validity of the study, since what is being claimed can be easily checked by an external eye, and they also ensure external validity since they try to go beyond the analysis of a single case and look at repeated patterns of use of a certain sequence. On the other hand, trying to categorise the phenomena in order to quantify the qualitative analysis may result in some loss of detail and richness of particulars.

The following chapters will try to start from qualitative and case-by-case analysis, making collections of similar occurrences of phenomena, and then to submit this analysis to quantification, which will provide material for some cautious generalisations about what happens in the data.

3
Simple Response Format to the Request

This chapter will start looking at the first of the different ways Rs provided responses to the requests made by Cs. As previously introduced, the analysis will be based on structural complexity and will look at the different formats of response used by Rs in the corpus. The first format of response presented is the less complex. No distinction was made between satisfaction and non-satisfaction of the request.

One kind of response to the request is a simple or basic response. This format consists of few turns and quickly leads to the end of the conversation. This kind of sequence is actually not very common in the calls under consideration (13 calls out of 141, i.e. 9.2 per cent). As will be pointed out in the different excerpts analysed, the occurrences of this pattern of response are mostly related to special encounters and they thus cannot be considered representative of the standard way of responding to service encounters. The first example, taken from the Italian calls, is in fact a call made to a pharmacy where the person picking up the phone is a child.

Example 1: Pharmacy, Sardinia (Italy)

((R is a child))

```
01  C:  ((telephone rings))
02  R:  >°pronto farmacia parenti°<     R: >°hello parenti pharmacy°<
03  C:  .hh e buongiorno mi scusi       C: .hh e good morning excuse me
        a che ora aprite:                  what time do you
        questo pomeriggio?                 open this afternoon?
04      ((R asks the mother in the
        shop))
```

05 R: aquattro e mezza   R: *half past four*
06 C: okey grazie   C: *okay thank you good[bye*
        arriveder[ci
07 R: [°>ar-rivederci <°   R: *[°>good-bye°<*
                        (SNI 16 *SS23F2 *SS FARMACIA)

The R of this call, being a child instead of the adult pharmacist, she needs to first check offline with her mother about the information requested before replying with the exact opening time. Then she simply says the time they will be opening the pharmacy that afternoon: a concise answer to the question, i.e. nothing less nor more than just requested. C then proceeds to initiate closing the call, as the response has fully satisfied her request, although formulated minimally (and not in a complete sentence such as 'we are opening at ...'). The format of this response can thus be justified because it is a child speaking who might not be accustomed to answering the telephone, so that she responds by just giving the relevant information but then does not yet have the competence for dealing with a formal telephone conversation. Moreover, as will be considered later on in the analysis of more complex formulations of responses, the initial delay of the response does not represent a feature of dispreference. This is so because in this kind of conversation both speakers seem to align and agree to a larger space for R to make the information requested available.

Some of the calls in our sample include requests for services. The following example contains one such request. In particular the call below provides a receipt for an order. These types of response also belong to this format of response to C's request, although this kind of call is somewhat different from the rest of the conversations in the corpus collected, in which C normally phones the service to ask for some information, i.e. they pre-request service. Nevertheless, this type of call represents a small proportion of the corpus (there are 9 calls out of 141, i.e. 6.4 per cent) and they are analysed together with the rest of the calls since their participants showed they could organise their responses in similar ways as when responding to pre-requests for service or general requests for information.

Example 2: Fruit and vegetables supplier, Cologne

01 C: ((telephone rings))
02 R: 'guten morgen?   R: *'good morning?*

03	C:	altmann guten morgen. ich rufe an von studentenheim junkersdorf,	C: altmann good morning. i'm calling from the junkersdorf hall of residence,
04		(.)	(.)
05	R:	jawoll	R: *yes*
06	C:	ich wollte auch die bestellung durchgeben für morgen (0.2) und zwar hätten wir gerne (2) einmal kiwi,	C: *i also wanted to give you the order for tomorrow (0.2) and we would like to have (2) one crate kiwi one of kiwi*
07	R:	**einmal kiwi**	**R: *one of kiwi***
08	C:	eine halbe kiste bananen	C: *half a crate of bananas*
09	R:	**mhhm**	**R: *mhhm***
10	C:	eine kiste granny smith	C: *a crate of granny smiths*
11	R:	**mh ja::**	**R: *mh ye::s***
12	C:	zehn kilo wirsing geschnitten	C: *ten kilos of sliced cabbage*
13		(2.2)	(2.2)
14	R:	**>hm<**	**R: *>hm<***
15	C:	einmal kartoffeln,	C: *one of potatoes*
16	R:	**ja:**	**R: *ye:s***
17	C:	einmal champignon	C: *one of mushrooms*
18	R:	**mh::**	**R: *mh::***
19	C:	einmal petersilie	C: *one of parsley*
20	R:	**mh::**	**R: *mh::***
21	C:	und zwei kilo sprossen	C: *and two kilos of sprouts*
22	R:	**mh::**	**R: *mh::***
23	C:	ja ich denk ah vielleicht noch eine kiste tomaten >wenn es geht<.ja?	C: *yes i think ah maybe another crate of tomatoes >if that's okay.< yeah?*
24	R:	**mh::**	**R: *mh::***
25	C:	gut das wäre dann alles	C: *good that's it.*
26	R:	**salat wollen sie nicht**	**R: *don't you want any lettuce***

27	C:	ne: salat haben wir noch	C: no: we still have lettuce
28		(.)	(.)
29	R:	jawohl.	R: yes
30	C:	ja danke schön	C: yes thank you very much
31	R:	danke::	R: thank you::
32	C:	wiederhören	C: goodbye
33	R:	wiederhören	R: goodbye

(CVD 14 *K50F *K Obst und Gemüse)

In this case R's main job during the call is that of acknowledging the items C is listing and showing he is taking note of it besides listening, whether through the repetition of the item requested (line 7), or by filling his conversational space with a continuer (*mh::*, and *yes*, lines 9, 11, 14, 16, 18, 20, 22, 24). The first *yes* R utters in line 04 (*jawoll*), even before C communicates the reason for calling, is to display R's readiness to receive the information from C and, as R's job is that of delivering fruit and vegetables, his readiness to take note of the order C is going to make. In fact, the order starts in line 05 without any check of the availability of R at that time, but just with the announcement that the order is going to follow. Both speakers seem thus to align to a preformatted type of conversation for the realisation of this call. The format also emerges in the initiation of the closing sequence through C signalling they are done (*gut das wäre dann alles*, good that would be all then, line 25), R checking she does not need anything else and followed by C's confirmation.

This type of call is like some others in the corpus with a request for an order or to make an appointment, where R provides the response in a simple format. This is just because, unlike in other calls in which C is ringing up to inquire about various services and products, here both speakers are certainly not making this kind of call for the first time. Rs of calls such as the one in the example above are accustomed to receiving these kinds of requests and the kind of response they give may follow routinised patterns. There is a similar familiarity on C's part that is never genuine to service inquiry calls. Moreover, the request for an order has a preformatted pattern that can be followed, which is given by the list in the order to be made. In this type of format R limits her/his answer to the information requested even if the request cannot be satisfied. Responses were considered minimal and displaying a preferred character even if they were marked by repeats, intensifiers, etc., but speakers reach the closing quite easily, as in the call below to the museum in Nuoro (Sardinia).

Example 3: Museum, Sardinia (Italy)

01	C:	((telephone rings))	
02	R:	pronto casa bianchi.	R: *hello casa bianchi (the Bianchi House)*
03	C:	buongiorno senta chiamo per avere un'informazione .hh volevo sapere se effettuate visite guidate per studenti di scolaresche anche al museo.	C: *good morning (listen)* R: *i'm calling to get some information .hh i wanted to know if you also do guided tours for student groups in the museum.*
04	**R:**	**sì sì certo.**	**R:** *yes yes certainly.*
05	C:	ah va benissimo. la ringrazio	C: *ah very good. thank you*
06	R:	prego arrive[derci	R: *you're welcome good[bye*
07	C:	[arrivederci buona giornata	C: *[good-bye have a good day*

(GDI 14 *NU31M2 *NU MUSEUM)

The museum is dedicated to a politician born in that part of Sardinia who worked in Rome at the beginning of the twentieth century. The museum is often visited by schools, as well as by tourist groups. It is therefore used to receiving calls asking for information about the services offered, which may be then followed up by an actual visit to the museum. Here the response comes after a yes/no question formulated in quite a detailed way in C's line 03, and R provides the information requested, marking it with the double repeat of *yes*, reinforced with *certo*, certainly. This example seems to show a general pattern in the way Rs respond to requests: as will be shown in the following chapters, R seems to feel to have to say something more than just answer the question that has been made (*sì, yes*). On the other hand, the feature that makes this call belong to this category is that the response is kept to its minimal meaning and just reinforced by the repetition, but there is no mention of other actions that may be implied in visiting the museum, such as booking for a large group, time and entrance fees.

3.1 The minimal format in the three languages

It was stated in Chapter 2 that this format is not widely used by speakers of the three languages considered here. Rs respond minimally in 9.2 per cent of the calls of the overall corpus ($N = 141$). There is actually another 14.9 per cent of calls in which R reacts to C's request with a minimal response and minimal response tokens, but they will be treated in another chapter (Chapter 6) because they occur with a specific sequential framework that needs to be talked about separately.

The calls with a minimal format of response described in this chapter are used to different degrees in the three languages of the corpus as shown in Table 3.1.

The figures presented in Table 3.1, as well as those that will be provided in the following chapters, provide a snapshot of the use of each response format and an indication of the different use of the formats analysed. In Chapter 7 then, the complete overview of the full corpus and the different formats of response used will be discussed.

Simply formulated responses thus seem to be used more often by the Italian Rs, representing more than half of the calls displaying this format, namely 14.3 per cent of the Italian sub-corpus. Germans and English chose this kind of response in 9.5 and 4 per cent of the calls respectively. In contrast to how this type of response has been formulated in the German and Italian samples, in English the instances in which R responds minimally never come with minimal response tokens as may happen in the other two languages, but with a full-sentence response.

Table 3.1 Simple response format*

Calls in the corpus	Simple response format (%)
English ($N = 50$)	4.0
German ($N = 42$)	9.5
Italian ($N = 49$)	14.3
Total corpus ($N = 141$)	9.2

* The values shown in the table are partial and they add up to those of the tables shown in Chapters 4–6. A table showing all formats of response used and their frequency can be found in Chapter 7.

4
Response plus Extension

Rs in the corpus seem to prefer to respond to C's requests in a more complex way rather than to restrict themselves to just giving the information they have been asked for, as in the examples described in Chapter 3. This chapter deals with responses that have been extended right after a first response has been provided. In particular a distinction will be made between:

- Responses extended only by apologies for not being able to provide the service requested.
- Responses extended through the addition of more information than is actually requested.
- Responses extended by the offer of a solution to the product requested.
- Responses extended through an account.

There may be two different types of account provided by R: one states that the service requested is not offered by the provider called; the other is produced to explain why the service called has or has not the product/service requested. The extension through an apology will be dealt with first.

4.1 Response plus apology

There are only few instances in which the response consists in providing the relevant information asked for by C, then followed by an apology, as the service cannot be provided. The next example displays this response format. It is an Italian call made to a coffee bar. The bar also sells sweets and cakes of different types and panettone (a typical Italian

46 *Business and Service Telephone Conversations*

Christmas treat) during the Christmas period, which is when the call was made. There is also a shop which forms part of the bar, where the above-mentioned products are sold together with chocolates and fine-quality sweets. Most of the requests received are requests for information, as described in Chapter 2.

Example 4: Coffee bar, Sardinia (Italy)

01	C:	((telephone rings))	
02	R:	pronto?	*R: hello?*
03	C:	pronto buongiorno vorrei avere un'informazione. avete del panettone all'ananas?	*C: hello good morning i'd like some information. do you have pineapple panettone?*
04		(0.2)	*(0.2)*
05	R:	un attimo che chiedo.	*R: just a moment i'll ask*
06	C:	grazie	*C: thanks*
07		((asking in the background))	
08	R:	no °mi dispiace°	*R: no °i'm sorry°*
09	C:	non ce l'avete? .hh la ringra[zio buongiorno	*C: you don't have it? .hh thank [you goodbye*
10	R:	[prego buongiorno	*R: [wel-come goodbye*

(CVI 14 *SS48F1 *SS caffè)

The response provided in this example is very similar to the responses provided by the Rs analysed in the previous chapter (Chapter 3), but here R does one action more than he did in the examples of Chapter 3: he provides the response to the request and apologises for not being able to provide the service. In this instance, R registers the request, gets the information from inside the shop and when he comes back to the phone, his response is short and polite, with no need to add anything to this, except for apologising for not being able to satisfy C's needs. At the other end of the line, C reacts by echoing the response, by which she signals the news has been received. She then proceeds to the closing part of the call by saying thank you for the information received and rings off. As Svennevig (2004) notes, cases of echo such as the one in the example above have a metalinguistic function:

> It does not commit the speaker to the truth of the previous statement [...], but merely shows that it has been registered. Such repeats generally occur after statements presenting new (and often

precise) information, and are therefore called 'information receipts'. (Svennevig, 2004, p. 490)

The repetition of the response as news receipt is quite a recurrent pattern in the corpus, as other examples will show. Apologies in this context are seldom used on the part of R. There are only three calls displaying the delivery of the response in a simple way and extended only by a routine apology in the corpus. One was English and two were Italian. In the analysis of the grammatical constituency of turn expansions, apologies were considered as free-standing increments, occurring either in the same turn of the core response, as in example 4, or in another turn. Rs opted more often to use other strategies to expand their response, as the next few paragraphs will show. The first that will be analysed is the addition of extra information to the response.

4.2 Response plus additional information

The first of the more elaborate types of response is the one in which we do get an answer to the request contiguously, i.e. in the turn following it, but this is then followed by an increment. The sequences present a continuation of the prior turn at a possible completion, sometimes in the same turn, and sometimes in a different one from the initial response.

The term 'increment' is drawn from Ford et al. (2002a), as introduced in Chapter 2, who define it as 'a non-main-clause continuation after a possible point of turn completion' (Ford et al., 2002a, p. 16). Ford et al. align with Schegloff's (1996) concept of extension, which is a continuation of the 'immediately prior possibly completed turn', and they also use this term, although they distinguish other continuations of the turn that belong to the category of 'increments'. So, while the extensions are '*constituents* of prior turn units' (Ford et al., 2002a, p. 17) in that they have a syntactic construction that continues the prior turn unit, there are also what they call *free constituents* among which they distinguish *unattached NPs* (term coined by Ono and Thompson, 1994):

> NPs that occur as increments after a place of possible completion but are *not* interpretable as syntactic constituents, or syntactically integrated continuations of the immediately prior turn. (Ford et al., 2002a, p. 17)

The next example is one in which the response is incremented after a point of possible completion and a slight overlap with the other speaker's talk. It is a call in Italian made to a museum to inquire about opening hours on Sundays. The museum is the same as in example 3 in Chapter 3, dedicated to an Italian politician. The request made is again within the range of requests dealt with by the museum staff over the phone.

Example 5: Museum, Sardinia (Italy)

```
01  C:  ((telephone rings))
02  R:  pronto casa bianchi.                    R: hello the bianchi house.
03  C:  eh:: buonasera (è) il museo?            C: eh:: good evening (is it)
                                                   the museum?
04  R:  sì?                                     R: yes?
05  C:  eh:: ascolti volevo sapere se la        C: eh:: (listen) i wanted
        domenica è aperto.                         to know if you're open on
                                                   sunday.
06  R:  sì sì è aperto tutti i giorni           R: yes yes it's open every
        (.) [eh::=                                 day (.) [eh::=
07  C:      [ah?                                C:         [ah?
08  R:  =eccetto il martedì                     R: =except tuesdays
09  C:  e ascolti mi può dire gli orari         C: erm (listen) could you
                                                   tell me the opening hours
```
(GDI 12 *NU40F1 *NU MUSEO)

In example 5 above, R is responding to a yes/no question, the preferred and default second-pair part of which would be responding just with 'yes, we are open on Sundays'. R actually seems to arrive at a first point of possible completion after she utters a complete clause, when she starts responding with extra information to the initial request, i.e. *it's open every day*. This first part of the turn is followed by a micro-pause after which overlapping speech by both speakers occurs: R retakes the floor to continue speaking and tells this to her interlocutor by using the continuer *e::*. C, on the other hand, shows surprise at the content of the response and makes relevant its possible completion with the token *ah?* R's addition of more information after the default response to the request has been delivered constitutes the next action, clearly marked by its occurrence after a PCP and the overlap with C's token.

In this case, the continuation of R's response belongs to the 'extensions' category, as it is syntactically linked to the first part; it is a constituent of the prior turn unit.

At other times, the first response is followed by an extension sequence that refers back to the request and asks for more details. The following call is an example of this kind of response followed by some increments, in which the increment is still part of the same turn. C is calling a restaurant to gather information about ready-made birthday party packages.

Example 6: Self-service restaurant, Frankfurt

01	C:	((telephone rings))	
02	R:	mittagtisch thiel katy grimm??	R: ((name of the restaurant + R's name and surname))
03	C:	ja >schönen guten tag mein name ist astrid hüber (.) ich habe eine frage< (.) und zwar würde ich ganz gern für einen geburtstag einen korb herrichten lassen?	C: yes >good morning my name is astrid hüber (.) i'd like to ask you< (.)that is i'd really like to have a basket prepared for a birthday
04	R:	hmhm	R: hmhm
05	C:	u:nd äh >hab ich da auch schon ne bestimmte vorstellung wenn ich da nen foto mitbringen würde können sie das dann ungefähr nachmachen?	C: a:nd eh >I already have an exact idea if i bring you a photo, would you then be able to make the same?
06	R:	wenn wir das haben	R: if we have it
07		(.)	(.)
08	C:	ja	C: yes
09	R:	können wir das sicherlich tun (.) für wann ist das?	R: we can do it for sure (.) when is it for?
10	C:	e:m das wäre für nächste woche ende nächste woche	C: e:m it would be for next week the end of next week
11	R:	ja kommen sie einfach mal vorbei und dann sprechen wir darüber	R: yes just come around and then we can talk about it
12	C:	ja. .hh e:m jetzt hab ich noch eine frage? ehm	C: yes. .hh e:m now i have another question? ehm

(ATD 05 *FFM30F *FFM TAGESBISTRO)

In this call R starts her response by projecting a conditional and possible situation (line 06, i.e. *wenn wir das haben, if we have it*) and gives the actual and explicit response in the first part of the following turn, *können wir das sicherlich tun* (*we can do it for sure*). As in the previous example, here the preferred second part of the request–response pair would be a yes/no answer. The format of the response is constrained by the form in which it has been projected in the first turn of the response. R decides to respond by formulating a complete sentence to show the ability to do the reproduction C is asking for. This type of response format is similar to those analysed by Lerner (1996), with a construction 'if/then', in which the 'if' (*wenn*) already projects and anticipates the second part is coming, the 'then' part (((*then*)) *we can do it for sure*). The last part of the turn is related to the response and asks for details of the business that has been requested, representing an attempt to make the call useful for both speakers: C receives the information she needs, and what she is looking for; R on her part, besides just giving some information, earns some money if she finds a new customer. The initial request for information made by C is here taken by R as a pre-request for service, so that she invites C to follow up with a visit to better discuss the service requested and eventually provide it (line 11).

The extension to the response comes as a new turn constructional unit that thus represents an example of a free constituent. It is a main clause not syntactically connected to the earlier part of the turn, and as in other instances it is separated from it by the occurrence of a micropause (line 09, *für wann ist das?*, *when is it for?*).

The last example in this section shows a different distribution of the increment through the sequence. In the next call, example 7, the response is extended in two or more different turns and comes after some feedback from C. The R of the call is an estate agent who mostly receives calls asking for information about the service but also prerequests for service that may be followed up with an upcoming visit from C. C's request here asks for general information on the different services offered.

Example 7: Estate agent's, Portsmouth (UK)

01 C: ((telephone rings))
02 R: thank you (   ) douglas susanne speaking how can i help?
03 C: .h good morning susanne. em i was em wondering e:m if you: also: (.) do things do with apartments (.) leasing apartments and thash

04   R: we do have a letting department
05   C: you do have a letting department.
06   R: yeah. >i can give you the number< it's actually a separate office [e::m
07   C: [°okay°
08   R: it's imp- it's a north end it's o two three nine two. [...]

(SNGB 13 *IR35F1 *POR ESTATE AGENT'S)

In the first response turn (line 04), R positively and literally responds to the query by confirming they have what C is looking for. C replies to this with a partial repeat to acknowledge the response. The repetition of the response by C with a slight variation is used here and elsewhere to fill a relevant place for turn transition and speaker change with fluent and automatic speech (Tannen, 1989). This is heard as an implicit go-ahead, by which C would be asking R to tell her more than just that they have the specified department (such as *I'll put you through to them*). Therefore the response token *yeah* at the beginning of line 06, confirms the response just given and repeated by C and treats this repetition as an implicit request for an extension of the prior turn with the relevant information. And this is what we get in the second turn of response (line 06), where R introduces the news, packing it into a free constituent extension: *>i can give you the number< it's actually a separate office.*

There is another option R can choose when responding to the requests for information, which is systematically offering an alternative solution to what has been asked for. This type of extension is analysed in more detail in the next section.

4.3   Response plus alternative solution

Another expansion of the response R can make is of the type of the next example, taken from the Italian corpus: when R does not have the precise item requested, s/he can offer an alternative item, whenever this is possible, of course. It is a conversation with an antique shop which is quite big. The products sold are of various kinds: they range from jewellery to small objects and to pieces of furniture. The shop mostly deals with customers face to face, but it occasionally receives calls asking for general information or whether it has specific items, as in the case below.

Example 8: Antique shop, Sardinia (Italy)

01	C:	((telephone rings))	
02	R:	pronto?	R: hello?
03	C:	pronto buongiorno	C: hello good morning
04	R:	buon[gio-	R: good[morn-
05	C:	[asco:lti un'informazione io sto cercando per un regalo >un servizio di bicchieri antico<	C: [(listen) some information i'm looking for a present >an antique set of glasses<
06	R:	e:h	R: e:h
07	C:	ne av[ete?	C: do you have [that?
08	R:	[antico no di bicchieri no .hh ho u:n bel servizio: di: >tazzine da caffè:<=	R: [antique no of glasses no .hh i have a nice se:t o:f >coffee cups<=
09	C:	=ah ecco=	C: =ah I see=
10	R:	=e:h insomma	R: =e:h well
11	C:	sono particolari [°comunque°	C: are they unique [°anyway°
12	R:	[beh è bavaria? >molto buona come qualità< poi è dorato con una <pietra rossa: sopra> (0.2) è da dodici (.) ci sono tre caffettie:re .hh e una lattiera	R: [well is it bavarian? >very good quality< then it's gilded with a <red stone on it> (0.2) it is for twelve (.) there are three coffee: pots .hh and a milk jug
13	C:	quanto vengo a spendere	C: how much does it cost
14	R:	e dunque quello costa sulle ottocento mila lire completo. da dodici completo eh? bellissimo ( ) bavaria è ottimo come hh. .hh qualità [...]	R: e well that one costs around eight hundred thousand lire for the complete set. for twelve eh? very nice ( ) bavarian is excellent hh. .hh quality [...]

(ATI 14 *SS23F3 *SS ANTIQUARIATO)

The call is one of the instances when a change of item can be made: C is looking for a present and has an idea of what to buy, *an antique set of glasses*. The present to be made gives R the chance to minimise the fact that she does not have a set of glasses and the opportunity to stress her new item as a solution to the search: a set, a nice antique set of coffee cups (see line 8). Moreover, the response comes in an initial overlap with the end of C's request, although at line 6, with *e:h*, R has already signalled her reception of the request and is bidding to get the floor. At line 10, then, after C's display of her orientation to the new item, R keeps giving details of the object she's selling over the phone now, and supplements it by giving details of its quality and a more precise description. This time R, after having suggested an alternative to C's request, focuses on the new item, and this is possible because C is receptive to the news. With *ah ecco* (*ah I see*), she responds to the presentation of the new item, displaying willingness to change her mind about the gift. This is also part of the reason why R needs more turns to try and leave his speaker with a 'satisfying' response to the request. In the present example the response stresses less the delivery of the bad news, R does not have what C has in mind, but more the alternative given. This results in a difference in the overall exchange of the turns with respect to instances in which R seems to orient to what precisely C is looking for and does this by questioning her/him, as will be shown in the following chapters.

This kind of increment of the response is considered as a free constituent. The alternative is introduced in the same turn as the response and then expanded with more details about the set after C's assessment request in line 11. The response in itself, on the other hand, comes with the double repetition of the features listed by C for the object required (*antico no di bicchieri no, antique no of glasses no*, line 08).

The next example, a German call to a furniture shop, shows the offer of another kind of solution instead: the directions to a place where to find the requested item. The types of requests that may be received in this shop are very similar to those of the Italian antique shop above. The difference lies in the fact that only furniture is sold.

Example 9: Furniture shop, Frankfurt

```
01  C:  ((telephone rings))
02  R:  (   ) guten tag?                    R: (   ) good morning?
03  C:  guten tag hier ist gerhard          C: good morning gerhard
        dietrich ich hab eine frage         dietrich speaking i have a
        ich hab heute ein geschenk          question i have (to buy) today
        für meine schwiegermutter           a present for my mother-in-
        nächste woche und (.) ich           law next week and (.) i know
        weiss dass sie so für fünfziger     she is into things of the fifties
        sechziger jahre sachen              and sixties=
        schwärmt=
```

04	R:	=[hmhm]	R: =[hmhm
05	C:	[u:nd wollte fragen haben sie so so nierentisch das war doch damals [so	C: [and i wanted to ask have you such such small tables it was certainly at that time [so
06	R:	[ja ja ne: habe ich nicht .hhh ((blowing his nose)) wer konnte das haben? .hhh ALSO DER EINZIGE LADEN der sehr gute sachen hat? ist der züglich? .hhh e::m ist den sachsenhausen in de::r (1.0) weil straße. gibt es eine kleinen laden. (0.2) der so speziell hm sachen mal (auf diese zeit) hat	R: [yes C: yes no: i don't have any .hhh ((blowing his nose)) who might have this? .hhh WELL THE ONLY SHOP that has very good things? is the (following)? .hhh e::m it is the sachsenhausen in (1) weil strasse. there is a small shop. (0.2) that has such special hm things (from those times)
07	C:	ach so. e::m haben sie denn (.) die a- e::m telefon nummer oder?	C: so. e::m do you then have (.) the a- e::m telephone number?
08	R:	ne ne ne ne ich kenne es ( )	R: no no no no i know it ( )
09	C:	das ist in welcher straße?	C: it is in which street?
10	R:	weil straße im sachsenhausen	R: weil strasse in sachsenhausen
01	C:	gut. dann schaue ich mal in ([ )	C: good. then i'll have a look in ([ )
12	R:	[bitte gerne	R: [you're welcome
13	C:	ja. okey? vielen [dank wiederhören	C: yes. okay? thank you [very much goodbye
14	R:	[tschüss	R: [bye

(ATD 14 *FFM22M *FFM MÖBELLADEN)

Both examples display those cases in which Rs turn bad news into good news with their responses (Schegloff, 2007). Here again, R displays her understanding of the request, following what C is saying with *hmhm* in line 04 in overlap with C's final part of the request. The dissatisfaction comes with an initial display of understanding of what C is looking for (*ja ja, yes yes*) and then with a direct response *ne: habe ich nicht* (*no: I don't have it*), immediately followed by R's self-questioning about where to find the requested things and the offer of another shop where C can go and find what he needs. This time again the response is incremented in the same turn through a free constituent. This call is also an example of a resource Rs can access while responding, that of their personal knowledge and experience; the R of the example above says this explicitly in line 08 (*ne ne ne ne ich kenne es, no no no no i know it*) to reply to a request that he cannot satisfy in his shop. This way of minimising the non-satisfied response can be read once again as a way of repairing the initial dissatisfaction, another way of satisfying the request.

The next section will deal with another way of extending the response, this time instead of adding more information or offering a solution to the request anyway, R chooses to give an account for offering or not offering the service requested by C.

4.4 R's offer of an account for the response

Calls in which the responses are justified are second-pair parts to firsts that have been formulated in a complex way and that are thus perceived as potentially problematic in asking for something specific and at the same time unusual. As for the extension through an increment adding information and giving alternative solutions, this type of extension is also made equally when the second-pair part does not meet the request and when it achieves satisfaction in the end.

R can provide at least two different types of account: s/he can either just state that the service requested is not among the offers provided, or give a more elaborate explanation about why the service requested cannot be provided. In the first instance R increments her/his response by conveying information that is wider than the request just made. In other words, by responding to a specific request for a service with a statement that includes the service requested and states that the overall service is of a different type than requested, R anticipates and avoids further requests on the same item or service. In the next example, for instance, R responds to a request for men's shoes by saying that the shop at which he works just deals with women's. This kind of request is possible since the only information available in the white pages were the name of the shop in the list of shoe shops. Also these kinds of increments of the responses are of the extension type, as they are usually produced as syntactically dependent on the initial response as the quantitative analysis at the end of the chapter will show.

Example 10: Boutique and shoe shop, Frankfurt

01	C:	((telephone rings))	
02	R:	von scheerbart?	*R: von scheerbart?*
03	C:	.hh guten tag hier ist gerhard dietrich ähm ich (.) folgendes problem ich hab ähm ich möchte eine wanderung machen und zwar (.) in griechenland und ich bräuchte da wanderschuhe dafür (.) ham sie sowas auch?	*C: .hh good morning this is gerhard dietrich ehm i (.) the following problem i hav ehm i'd like to go hiking specifically (.) in greece and i need hiking boots for that (.) do you have something like that?*
04	R:	**für männer hab ich- wir gar keine schuhe (.) nur für frauen**	***R: for men i- we have no shoes at all (.) just for women***
05	C:	ach? sie haben nur frauen schuhe?	*C: ah? you just have women's shoes?*
06	R:	ja:	*R: yes:*
07		(.)	*(.)*
08	C:	achso. (okey dann) weiss bescheid	*C: i see. (okay then) i see*
09	R:	gu[t	*R: go[od*
10	C:	[danke schön? (0.2) wieder hören	*C: [thank you very much (0.2) good bye*

(ATD 06 *FFM22M *FFM SCHUHGESCHÄFT)

In this example C explains he needs hiking boots for going hiking in Greece and R responds with a shift from the first person singular to the first person plural (*I* versus *we*). The response comes with a very brief formulation when compared to other examples such as the previous one, example 9; R gives the relevant information plus a brief explanation for not having men's shoes (they just sell women's shoes), without spending much time for doing so: the request cannot be satisfied and besides, more talk is not relevant as R cannot help C with his request at all. The result of the response is also made relevant in the short reply (*ja:, yes,* line 06) to C's reaction to the news with surprise (*ach*) and a repeat of the information received.

Repetition plays a relevant role when news are received, through its formulaic recycling of previous talk from the interlocutor, speakers take time to think about what to do next. At the same time, here the repetition works as an implicit go-ahead request to the other speaker. In the previous example repetition does not seem to work either way: it seems not to give enough time to C to project a new action for obtaining the service required, nor does it prompt R to suggest a possible solution. Rather R shows the completion of his response by the minimal response token *ja* (*yeah,* line 06) to C's second request asking for confirmation of the news.

There are also instances in which the response is followed by an account of the second type, in which an explanation for not having the required service is provided. Such a response can also last over more than one turn and also constitute a long sequence, as in example 11 below, a call to a bookshop chain in Portsmouth. The call was made at a time when a Harry Potter book was announced to soon be appearing in bookstores.

Example 11: Bookshop, Portsmouth (UK)

```
01   C: ((telephone rings))
02   R: °good afternoon all^books?°
03   C: .hh oh good afternoon i'm interested in the new harry
     potter book .hh i wasn- are you taking orders for (it)(so) when
     does it c:ome out.
04   R: .hh there's still no publication date i'm afraid (0.5) a:nd
     so i don't think °we can actually place an order 'cause then
     we won't be able to give an isbn°.
05   C: i see . so i have to leave it a while.
06   R: yeah. [( )
07   C:      [okay.
08   R: i'm after it myself but? [you know=
09   C:                          [yeah
10   R: =we- i think (0.2) leave it another mmonth (0.2) and
     there we would-should be able-we-as soon as e: they give us
     a publication date they'll give an isbn number as well?
```

11 C: okay
12 R: and then we can place °your order°.
13 C: thank you very much
14 R: okay?
15 C: yeah
16 R: [°thank you°
17 C: [bye
18 R: bye

(GDGB 1 *POR50M1 *POR BOOK SHOP)

In this instance the response is made with a first repairing sequence where R explains and completes the response (.hh there's still no publication date I'm afraid, line 04). This may be compared to the previous example in which in the first instance R pointed out the impossibility of providing the service at all. By saying a publication date has not been decided on yet, R implies that she cannot do anything, but increments this first response by providing an apology (I'm afraid), the 0.5 second gap within the turn and the stretch of the beginning of the next sentence (a:nd so) in the same turn.

Once again the answerer initially speaks in the first person (i don't think), minimising the weight of the negative response, then she starts talking on behalf of the company and lowers her voice (I don't think °we can actually place an order cause then we won't be able to give an isbn°.). C, in his next turn, collaborates with the completion of the response by bringing out the implications of the news received: if they cannot place an order, he thus has to wait for a while (line 5: i see. so i have to leave it a while.). C's next turn also points to the closing part of the call, through the free-standing okay. But the closure is postponed at R's next turn, where she initiates another repair sequence from lines 08 to 12, by giving further accounts, first justifying her response with her own interest in buying the book, and then trying to leave her interlocutor with the most positive response (as soon as e: they give us a publication date they'll give an isbn number as well). The rest of the turn (line 10), after the pre-sequence just pointed out with R's personal interest, is also reformulated while she speaks: there are two pauses within the speech flow in line 10, the self-correction in the choice of the verb used, and the self-correction in reformulating the concept she wants to express (we- i think (0.2) leave it another mmonth (0.2) and there we would-should be able-we-as soon as e: they give us a publication date they'll give an isbn number as well?).

This repetition of news, once it has already been delivered, seems to point out two issues:

- It works as a further repair, as just said, for not being able to satisfy the request;
- And displays the speaker's orientation to avoiding abrupt and minimal responses.

In these cases what seems to be followed is a rule for 'satisfaction', i.e. the tendency to always give a response that meets the customer's needs, and if this is not possible, the speaker makes an attempt to minimise it. And by doing this, they actually utter more complicated response turns that take a dispreferred form.

4.5 Response plus extension: more complex responses

Sometimes the responses in the corpus were incremented by the use of more than one of the actions described in the previous pages. R can increment the response by giving both types of account described, or by giving an account and following this by, for instance, the offer of an alternative solution. The next example is an English call to a bookshop and contains strategies of response that can be considered normal for bookshop assistants in Great Britain, but probably also everywhere R has received some training in communication with the client both over the telephone and face to face. This is because English Rs seem to have had relatively more training than their colleagues in Italy and Germany. Although all the calls are directed to shops and services of on average the same size, in Great Britain they are mostly employed in commercial chains that dedicate more attention to customer service over the phone.

In these calls there seems to be a precise order for providing the response. The first move provides an account for not being able to grant the request immediately.

Example 12: Bookshop, Portsmouth (UK)

01   C: ((telephone rings))
02   R: good afternoon all^book(s). can i help you?
03   C: yeah. i'm just making an enquiry do you have hm (.) i don't know if it's a biography or an autobiography for victoria beckham? .hh do you have it in st[ock?

04 R: [no no we don't stock ehm general biographies unfortunately.
05 C: right.
06 R: (we could order) to you one. e:m do you want it today?
07 C: .hh no it's i- i- i- it's not necessary, no. [i were=
08 R: [no
09 C: =just (.) doing an enquiry.
10 R: okay. no. we don't actually have one in stock, but i can certainly research one for you ( )
11 C: e:m i've got- i've got a hurry at the moment. but e:m i'll phone you back later. [°okay?°
12 R: [okay a: well we're closing in fifteen minutes.
13 C: okay.
14 R: okay? yeah? e we do have a web site if you have got access to the internet?
15 C: okay, what's your website?
16 R: it's double u double u double u dot
17 C: yeah?
18 R: allbooks? [dot=
19 C: [yeah?
20 R: =co dot uk.
21 C: okay.
22 R: all right?
23 C: thank you. bye?
24 R: bye-bye

(GDGB 10 *POR29M1 *POR BOOK SHOP)

This call is directed to the same bookshop as in example 11 above. In line 04 R begins his turn by giving first the response to the query with the double repetition of the response token no: *no no*, and then increments the delivery of the bad news by giving a first account for this *we don't stock ehm general biographies*. The third step of the response is now an apology, expressed with *unfortunately here*. The fourth step then represents the offer of a solution, made here with the suggestion to make an order for the book, coming in the following turn (line 06). This move is made after C's receipt token *right* that signals the information has been registered and a change of state has occurred: the item requested is not available. Nevertheless the offer is rejected by C, which minimises the importance of finding the book at that precise moment. R accepts the rejection with his *okay* in line 10 and by repeating his response. This time R's double repetition of the non-satisfied response marks the impossibility of doing anything else but offering what he is saying. In the last part of the call (line 11 onwards) the speakers switch to another question, that of getting the book anyway, despite the hurry displayed by C, and R first points out that it's 15 minutes before closing time and then he suggests another solution, that of checking on the website. This part of the conversation is also characterised by the production of *okays* in each speaker's turn, signalling the reception of relevant bits of information as well as working as pre-closure markers in search of a satisfactory one for the call. By doing this kind of work, *okays* keep their nature of closure-relevant and continuative tokens at the same time, as they are relevant both for prior and next-positioned matters (Beach, 1993). They thus work similarly to the change of state token 'oh' analysed by Heritage (1984a).

Examples analysed in this chapter are quite different from those taken into consideration in the previous one (Chapter 3), and show one possible way of extending the response turn by complicating its structure. This also responds to the key features for dispreferred response formats as described in Chapter 1. The next section will observe a special case, an odd example of R's increment of the response.

4.6 Extension to the response: an odd example

The last example to be discussed in this chapter is an excerpt in which the response to the request is also extended, but R also seems very chatty, so that it is him/her leading the conversation and keeping the channel open, as in the call below to a travel agent's in Italy.

Example 13: Travel agent's, Sardinia (Italy)

01	C:	((telephone rings))	
02	R:	>sì pronto?<	R: >yes hello?<
03	C:	hh e pronto buonasera è il centro soggiorni studi?	C: hh a hello good evening is that the centro soggiorni studi?
04	R:	sì buonasera mi di[ca	R: yes good evening how can [i help (tell me)

05	C:	[.hh e: buonasera ascolti ho saputo che fate corsi di inglese: e spagnolo .hh[h	C: [.hh a: good evening (listen) i found out that you do english and spanish courses.hh[h
06	R:	[sì	R: [yes
07	C:	ecco volevo avere alcune: informazioni.	C: well i wanted some: information.
08	R:	sì	R: yes
09	C:	verso: -magari mm. che ora posso venire? .hh m: con chi posso parlare ecco?	C: at around – maybe mm. what time can i come? .hh m: and who can i talk to?
10	R:	ma ss- guardi >se vuol venir siamo benissimo qua< sicuramente fino alle otto almeno. [quindi=	R: well ss- (look)>if you want to come we are here< certainly till eight at least. [so=
11	C:	[ah	C: [ah
12	R:	=se vuole venire? se vuole fare un salto? ci sono io? sono: io mi chiamo rosario, sono: responsabi[le per quanto riguarda il settore=	R: = if you want to come? If you want to pop in? there is me? i:'m my name is rosario, i'm responsi[ble for the sector=
13	C:	[ah?	C: [ah?
14	R:	=corsi di lingua e viaggi studio	R: language courses and study trips
15	C:	ho capito	C: i see
16	R:	quindi? se vuole venire anche adesso: sono qua? [insomma.	R: so? if you want to come even no:w well i'm [here?
17	C:	[va bene. magari mi avvicino [stasera o:::	C: [okay. maybe i'll come [tonight or:::
18	R:	[sì sa dove siamo a::l via la marmora? presso il centro commerciale >acqua e te<. dove c'è l'arco praticamente.	R: [yes do you know where we are a::t via la marmora? in the shopping centre >acqua e te<. where the arco is in fact.
19	C:	ahah ho capito. va [bene	C: ahah i see. o[kay
20	R:	[d'accordo?	R: [all right?

21	C:	grazie
22	R:	oppure domani mattina dopo le undici e mezza
23	C:	sì?
24	R:	ci sono anche all'una e un quarto. sicuramente ci sono. ma? se può venire stasera? forse è meglio così cominciamo a vedere qualcosa anche perché .hh stiamo ultimando le iscrizioni per il (.) semestre adesso
25	C:	ah
26	R:	e quindi: [occorrerebbe
27	C:	[perché adesso inizia un nuovo corso
28	R:	sì. praticamente: stiamo facendo gli ultimi inserimenti adesso .hh
29	C:	ahah
30	R:	quindi:: insomma, ecco
31	C:	e-ecco pri:ma:: vengo meglio è insomma.
32	R:	forse sì. è meglio.
33	C:	va [bene=
34	R:	[d'accordo?
35	C:	=la ringrazio
36	R:	grazie a lei ar[rivederci
37	C:	[arrivederci

C: thank you
R: or tomorrow morning after half past eleven
C: yes?
R: i'll be in also at quarter past one. i'll be in for sure. but? if you can come tonight? maybe it's better this way we can start looking at something also because .hh we're closing the enrolment for the (.) semester now
C: ah
R: and so: [it would be necessary
C: [because a new course is starting now
R: yes. in fact we're registering the last names now .hh
C: ahah
R: so:: well, that's it
C: so-so then the earlier i come the better
R: maybe yes. it's better.
C: o[kay=
R: [all right?
C: =thank you
R: thank you go[odbye
C: [goodbye
(SNI 05 *SS CVS)

The travel agent's deals specifically with organised trips for students, i.e. study trips abroad. It thus arranges not just the journey and the accommodation, but also language courses in the different countries as well as at the centre. The calls it may receive are of three types: those asking about study trips as in the call under examination, those asking about language courses held at the centre, and those asking for general information.

C's request here is quite simple, and at the same time twofold. After a pre-sequence in which she introduces the reason for calling and for making her request, the fact is that she just wants to know if the travel agent's also offers language courses, and she wants to know the opening times of the place and the person she can talk to for this service. The response could thus simply come as in other calls analysed before in which R lists the opening times and adds to this the name of the person C can look for once she gets to the centre. But R just opts for a different style of response and starts off by giving the closing time for that evening and repeating three times the offer to visit the centre as early as that afternoon (*se vuol venire, if you want to come*, lines 10–12). After this R starts responding to the second request regarding the reference person by introducing himself as 'me' in the first instance, repaired in the next few seconds through the introduction of the interlocutor by name. This is then expanded by the addition of more information about his duties, thus giving weight to the self-reference. This first increment of the response comes as syntactically independent from the rest of the response, and is thus treated as a free constituent. The call does not lead to an end here, as the transcript shows, but both speakers do seem to orient to a closing at this point, after C shows the information has been received in line 15. Lines 16–21 show two attempts on C's part to close the conversation through explicit signs of satisfaction about the news received (*ho capito, va bene, I see, ok*) and reference to an upcoming visit to the travel agent's, after R repeats for a fourth time the offer of his availability to help his speaker once she visits the centre, even immediately (*anche adesso, even now*). C's first attempt to close the conversation is postponed by R's further expansion of the response with the addition of directions for reaching the centre, in overlap with C's talk. C's indication that she will visit the travel agent's is sufficient for R to start giving directions (line 18, *sì sa dove siamo a::l via la marmora? presso il centro commerciale >la marmora<. dove c'è l'arco, yes do you know where we are a::t via la marmora? At the shopping centre >la marmora<. where the arco is in fact.*) even before C actually specifies when she wants to come.

But what is more striking in this conversation is how the exchange develops in the following turns (lines 19–21), in which both speakers seem first to overtly orient to the closing through assessment and thanking, and they then turn back to full discussion in line 22 with R continuing the conversation and offering another possibility for C to follow up with a visit by providing subsequent details of opening times. This is then followed by an orientation back to the first offer of a visit, this time supported by an account (i.e. line 24, *forse è meglio così cominciamo a vedere qualcosa anche perché .hh stiamo ultimando le iscrizioni per il (.) semestre adesso*, maybe it's better this way we can start looking at something also because .hh we're closing the enrolment for the (.) semester now, and so on).

The main issue in this excerpt is the large expansion on R's part compared to the mean of the other calls. What seems to distinguish the response in this call is not the fact that the response is expanded, but that it seems to take more time than what is actually necessary to respond to such a simple request. The length of the conversation is not justified, as in other instances, by the search for an acceptable solution to a complicated issue. What seems to play a major role is R's training to be nice, although he ends up by just being chatty.

4.7 Cross-cultural comparison: response plus extension

In this chapter we have seen a general pattern of response to the requests that may be performed in different ways depending on the specific request each time. The extension of the request through an apology, or an increment that adds more information to the response, either irrelevant or relevant because it serves to repair the non-availability of a service, is used in 35.4 per cent of the calls. The four types of turn extension in this format are distributed as follows:

- 2.1 per cent of the calls in which R extends the response with an apology;
- 7.1 per cent of the calls in which R provides additional information in the response to the request;
- 7.8 per cent of the calls in which R gives an alternative solution to what is being asked for;
- 11.4 per cent consists of calls in which the extension to the response is an account, a statement that clarifies that the actual services provided are different from those requested or a justification for why R deals with that type of service, why it is good or bad, etc.;
- And another 7 per cent of the calls consists of conversations in which the extension of the response is more articulated and includes more than one of the above actions extending the response.

The kinds of actions performed by the expansion of the initial response show that an apology as only resource to repair the lack of service is seldom used. Instead, other ways of extending the response are used much more, i.e. those that comply or do not comply with the request. This stresses the existence of at least four possibilities for the speakers of the three languages to expand their responses, among which three are more used: either by giving more information about the request, or offering an alternative solution to it, or justifying the response. Then there is a fifth option consisting in performing at least two of these actions together, often combining the offer of one type of account together with the offer of a solution, or both types of accounts, an apology, and the offer of a solution as a final resource.

Speakers of all three languages seem to prefer one kind of extension more than the other, namely the display of an account. Table 4.1 shows the results of the cross-cultural comparison for the use of this format of response in the three languages.

Of the calls displaying the format of response analysed for this chapter, English speakers seem to extend their responses more often; they do this 44 per cent of the time, followed by the Italians with 36.6 per cent. Germans, on the other hand, do this less often, in 23.9 per cent of the cases.

Table 4.1 shows that among the options for incrementing the response, the English seem to prefer not just to perform one action. Rather they provided an alternative solution in 12 per cent of the calls, the same percentage for which they combined the various options and mitigated the news in more complex ways (12 per cent). They produced an account of the response in 10 per cent of the calls in the corpus, and provided additional information with respect to what was requested in 8 per cent of the calls. English speakers provided an apology in 8 calls out of 23, 7 of which were produced together with another increment and one as the only mitigating strategy (2 per cent).

Moreover, Germans increment their responses less often than their colleagues in Great Britain and Italy, and they never chose to extend the response turn by adding more information or apologising; only 4.8 per cent of the time did they give an alternative solution, or combined this together with providing an account in another 7.2 per cent of the calls. Almost 12 per cent of the calls in German were registered with the only choice of an explanation and justification for having or not having that kind of product being the most used response strategy in this format.

Finally, Italian speakers extended their turns using all the options described for turn extension, again having a preference, just like the Germans, for a justification of the response (12.2 per cent). The addition of more information about the service requested was also used, more

Table 4.1 Extensions of the response (%)*

Type of response extension	English (N = 50)	German (N = 42)	Italian (N = 49)	Total corpus (N = 141)
Apology	2.0	0	4.1	2.1
More information	8.0	0	12.2	7.1
Alternative solution	12.0	4.8	6.1	7.8
Account[†]	10.0	11.9	12.2	11.4
Combined actions[‡]	12.0	7.2	2.0	7.0
Response plus extension total	44.0 (N = 22)	23.9 (N = 10)	36.6 (N = 18)	35.4

* The values shown in the table are partial and they add up to those of the tables shown in Chapters 5 and 6. A table showing all types of increments used in the cross-cultural corpus and their frequency can be found in Chapter 7.
[†] The calls included in this calculation show the production of only one type of account in extension to the response, i.e. either a general statement that displays the service deals with something different than requested, or a more elaborate explanation in which R justifies why s/he does not provide the service asked for.
[‡] These responses provide the offer of a solution after a display of an account in English and German. The Italian call (there is only one call displaying this format) instead displays both types of account complementing one another.

often than in English, in 12.2 per cent of the time. The offer of a solution was used in 6.1 per cent of the time alone, plus in another 2 per cent it is provided together with an account. Apologies, as mentioned above, were also found only as unique mitigating strategies in 4.1 per cent of the calls.

These results show a general tendency for the speakers of the three languages to increment their responses through the use of an account. But the results also outline possible differences in the preference for incrementing the response by adding new information about the service requested in the Italian and English calls compared to the German ones. And other possible differences among the three languages are given by the British offering an alternative solution to the request more often than the speakers of the other two languages.

4.8 Summary of features: response plus extension

This chapter has dealt with one format of response to requests in service encounters that contributes to outlining the features of second-pair parts in this particular setting. The extension of the response has been described as a feature for dispreference, as it is a way of complicating the structural formulation of the response. In the relevant literature,

Levinson (1983, p. 334) included in the features for dispreferred parts only one of the extensions used by the speakers in the data analysed here: the formulation of accounts. The analysis of bookshop service encounters also found that the shop assistants volunteered an alternative solution to the book requested (Pixi Project, Aston, 1988a; Zorzi, 1990; Brodine, 1991). The analysis has pointed out another way of making the response take a dispreferred character, namely by providing extra information about the service required, although this has not been requested.

Dispreferred second-pair parts thus seem to be accompanied by more talk. Purposeful silences and disfluences occur, but do not necessarily seem to be a premonitory sign of the forthcoming dispreferred sequences. In this particular communicative event, silences occurring before the response is uttered seem mostly to mean the absence of the available information requested by C right at the moment at which the request has been formulated.

Finally, the extension of the response can occur either in the same or in more than one turn, after C has displayed some kind of receipt to the news, and the type of increment can change. Tables 4.2 and 4.3 show the frequencies of occurrence of the different types of increments depending on whether the increment is initiated in the same or in a

Table 4.2 Types of increments of the response: the entire corpus (%)

Type of increment	Free increment	Extension/embedded increment	Total ($N = 141$)
Total ($N = 50$)	76.0 ($N = 38$)	24.0 ($N = 12$)	35.4

Table 4.3 Types of increments of the response: cross-cultural overview (%)*

Type of increment ($N = 141$)	Free increment		Embedded increment	Extension		Total
	Same turn	Other turn		Same turn	Other turn	
English ($N = 50$)	12.0	18.0	2.0	12.0	–	44.0
German ($N = 42$)	7.2	14.3	–	2.4	–	23.9
Italian ($N = 49$)	20.4	8.2	–	8.2	–	36.8

* The values shown in the table are partial and they add up to those of the tables shown in Chapters 5 and 6. A table showing all types of increments used in the cross-cultural corpus and their frequency can be found in Chapter 7.

Response plus Extension 69

later turn after the initial response, giving respectively the overall tendency in the corpus, and comparing the three languages and subcorpora analysed in this book.

Tables 4.2 and 4.3 show that free constituents are used more often than syntactically bounded extensions. Rs expand their responses right after possible points of turn transition in which C may produce an assessment or some receipt of the news just received, either with a receipt token (*okay, ah?, oh? yes, no*, etc.), or by the repetition of the news, as shown in some of the excerpts (cf. examples 7 and 10). If Cs then interrupt the conversation flow to display a receipt, free constituents incrementing the previously initiated response occur in a second turn after R's initial response. This apparently happens more often in English and German, accounting for 18 and 14.3 per cent of the calls in the corpus respectively (cf. example 12). Italian Rs, on the other hand, seem to extend their response through an increment that is syntactically independent from the previous response part in the same turn unit more often than in separate turns, 20.4 per cent, compared to constituents produced in another turn, 8.2 per cent. This different pattern in the Italian data can be explained by looking at the instances in which the free constituent occurs in the same turn: they are all instances in which R does not reach a TRP until he has finished delivering the information, packed in long turns. Moreover, half of the calls showing this type of expansion (5 out of 10) show a regular pattern at the beginning of the turn, marked by the use of discourse markers typical of spoken Italian, such as *guardi* (literally: *look*). A similar discourse marker used at the beginning of a TCU has been found in Swedish conversations playing the same role, the summons signal *hördu* (*listen, you know*): 'This summons signal, which contains the second-person singular pronoun *du*, is usually attached to a TCU rather than constructed as an own TCU in a presequence utterance' (Lindström, 2006, p. 88). These are some of the few cases in which R seems very chatty and is pleased to have the opportunity to show his products and services to somebody, as in examples 8 and 13.

The extension syntactically bound to the first unit of response occurs less frequently in this part of the corpus, accounting for 8.5 per cent in the three languages altogether (see examples 5 and 10). In contrast to free constituents, all these extensions come in the same turn unit of the initial response. Again, they are mostly used by British speakers, 12 per cent, followed this time by Italians in 8.2 per cent of the cases, whereas Germans seldom used this type of extension, in only one call, accounting for 2 per cent of these calls.

Finally Table 4.3 also gives an account of calls in which R embeds the extension in the response, directly providing an alternative solution to the request. This is only one instance, a call to a bookshop in the English corpus that comes after C has been put on hold for half a minute while R searches for the book requested. It is an exceptional case that could probably only happen again in the same circumstances, after the conversation has been suspended, but this needs to be observed in more data.

The next chapter (5) will try to deal with another feature of the way Rs provided their responses to requests, by looking precisely at responses that are delayed before actually being provided.

5
Insertion Sequence Followed by the Response

The response formats analysed so far display growth in structural complexity, as announced in the introduction to the analysis. Simple and expanded response turns observed in Chapters 3 and 4 respectively can represent, so to speak, the basic sequential developments in which responses to requests in service encounters can occur. The sequences analysed from this point on are structurally more complex because they bring into the conversation new sequences, but they also embed and combine the formats of response already analysed.

The third type of sequence to observe is one in which the response is delayed by R's initiation of an insertion sequence. The reasons for initiating this sequence instead of directly starting to respond to the request may be various and mostly take the shape of a request for more detail or a quick confirmation check. Among the circumstances in which these sequences occur, there are instances such as the following:

- R asks for confirmation about what C has just requested.
- R asks for repetition of the request because s/he has forgotten it, often when the call has been put on hold because it was interrupting an ongoing activity that could not be postponed.
- R completes the request initiated by C by asking for more details through yes/no questions.
- R narrows down what has been requested with a series of follow-up questions for details, because the initial request is too general for response in just one turn.
- R has not understood the request very well and asks for clarification, initiating a repair sequence.

After this sequence the actual response is given, occurring in formats similar to those described in the previous chapters, i.e. through

72 Business and Service Telephone Conversations

a simple or an expanded turn. In all cases this format is accomplished with more than one turn unit and displays more complex structures than those observed so far. The suggestion here is that not just calls receiving negative responses after the insertion sequences are produced in a dispreferred format, but all calls with such formats of response are dispreferred, i.e. also calls where R satisfies the request made by C, because they display much more complex structures than simple response formats. The analysis has shown that most of the calls in which an insertion sequence is initiated receive positive responses and provide the service requested, indicating a lack of information provided by C in the initial request, and the continuous need for both speakers to manage and arrange the information exchanged in the conversation. But what matters for the purpose of this study in the way these calls are responded to is their structural complexity.

The chapter looks first at single cases as representatives of what has been observed in the corpus, following the different circumstances that have caused the initiation of the insertion sequence. The analysis will also focus on the way in which the response is formulated and the adjacency pair completed. Finally, similarly to the previous chapters, the scope is extended towards the end of the chapter, looking at cross-cultural comparison and the general trend followed by all calls in the corpus.

5.1 Confirmation check as insertion sequence before the response

As mentioned at the beginning of the chapter, sometimes the insertion sequence initiated by R can just be a confirmation check about the request just heard, as in the example below (example 14). The call is made to the paper museum in a small English town. As for other calls to museums in other countries, here too the kind of requests received mostly involve information seeking.

Example 14: Museum, Lancaster (UK)

```
01   C:   ((telephone rings))
02   R:   good afternoon the paper museum?
03   C:   .hhh hello i was wondering if you could tell me e:m what
          are the opening hours fo:r the weekend for the mus-
          museum
04   R:   ehm (.) the paper museum?
```

```
05  C:  .h yeah
06  R:  e: we're open everyday ten till five apart that we're not
        open on a sunday
07  C:  .hh not open sun[days .hh e:m
08  R:                   [mh mh
09  C:  okay. and will the ehm times change at all for the christmas
        period? [or it stays the same?
10  R:          [eh: no: yeah we're closed over e:m i'll just just
                double check right now hang on and i'll tell you when
                we're [are closed
11  C:            [okay thanks
12  [...]
```

(CVGB 17 *LON33F1 *LCS MUSEUM 1)

In this case the confirmation check constrains the environment of the request, although this does not seem to be relevant: R already self-identifies with the name of the museum, the Paper Museum, and not just with the museum's category. This kind of insertion sequence seems to be mostly used by the British Rs: they used confirmation checks following the request in 7 calls out of a total of 12 with this kind of sequence preceding the response.

The request–response pair sequence in this excerpt is completed in line 06 after confirmation has been sought and obtained, by giving general information about the opening hours instead of just those for the weekend as requested by C (line 03). C, then, makes relevant the information she was looking for through a repeat of the relevant bit of information (line 07). This repetition prompts a confirmation from R that occurs here in overlap with the final part of C's repeat through the use of a weak acknowledgement token, *mh mh*, thus leaving space for C, who is recognised as the current speaker and who bids for a continuation of her turn by breathing in, then following with the production of a continuer (line 07). The repetition of the news also gives C time to move to a second, more specific request, about the opening times during the Christmas period. The response here comes expanded before the actual response is given, with general information embedded in the rest of the response (line 10).

The confirmation check sequence often consists of a partial repeat of C's request, and this may be heard as a way of making more relevant the issue of the request and the extent to which it is satisfied (the museum is not open on Sundays but it is on Saturdays).

Sometimes Rs can thus delay their response because they first ask for confirmation of the request. This is an additional move they make when they respond, as can be noticed in example 15 below, where, after initially asking for confirmation, R replies by explaining the reason why she refuses or does not advise C to request the service that way.

Example 15: Travel agency, Portsmouth (UK)

```
01  C:  ((telephone rings))
02  R:  good runners, good afternoon [this is](   )?
03  C:                               [.hhh ]
04      ((noise))
05  C:  >oh hallo there.< ahm (.) just one thing. can i book a
        ryanair flight in your agency? at all
06  R:  ryana:ir?=
07  C:  = yeah >do you do< with ry-ryanair.
08  R:  well they've direct service (their) ticket-desk. You (.) do
        it over the internet eh with the credit ca:rd(s).=
09  C:  = right okay.
10  R:  i mean we can do it for you. using ou:r system. But it
        means we're charging you=
11  C:  okay
12  R:  = (  ) twenty five pounds a ticket just for doing it
13  C:  okay.-okay.(.) so it's better on the internet.
14  R:  well, it's better for you:.
15  C:  yeah
16  R:  you get a better fare
17  C:  .hh okay. well, thanks for your honesty. thank you.
18  R:  ((laughs)) ahh. okay?
19  C:  okay. thanks a lot. by:e?
20  R:  °by:e°.
```

(GDGB 18 *POR50M1 *POR TRAVEL AGENCY)

Here the call is directed to a travel agency, and R may be prepared to receive more general questions as well as specific ones, such as an actual request for a service, a booking for instance. The agency, as most of the travel agencies at the time of recording, deals mainly with most companies without any extra charge, but applies one when having to book journeys with low-cost companies. At time of recordings though, low-cost companies such as Ryanair were starting to expand, and most people who had heard of them for the first time were not sure about whether they were able to book a cheaper flight from the usual travel agent or not.

R's confirmation check in this instance is first confirmed by C, who also restates the request immediately afterwards in the same turn unit. The response comes in the following turn (line 08) and R first gives an explanation about how Ryanair normally works, and targets the response in an indirect way. The statement about the normal functioning of Ryanair's booking system is followed by a second part of the response in the same turn: R tells her speaker what he should do and she thus offers a solution (*you (.) do it over the internet eh with the credit ca:rd(s).*, line 08). In the following turn, after C's displaying understanding and reception of the news, R continues by explaining the reason why she is sending C to check on the internet and refusing to do it for him, although she could. So this time R is representing C's interests (*well, it's better for you:., you get a better fare*, lines 14–16), and this is appreciated by C at the end of the call, when he thanks R for her honesty (line 17: *.hh okay. well, thanks for your honesty. thank you.*).

As in example 14 above, the response to the request after the insertion sequence comes expanded by an account, started before the response and continued in the following turn, after C's initial reaction to the news, but is also complemented by the offer of a solution, displaying a format of response in which more strategies are combined in order to minimise a response that does not grant the request. All the expansions here are uttered as syntactically independent units; they thus belong to the free constituents' category. This type of increment is produced as an afterthought, not initially projected together with the response provided in line 08, but made relevant by C's production of the assessment in response to the information delivered so far with *right okay* and not followed by subsequent talk by C (line 09).

In the third example for this chapter (example 16), R's asking for confirmation allows him to start checking whether the person searched for is already in or not. This call is actually made to the porter's lodge of a music school, so R's main job is usually that of checking if people are in or out, and putting through the calls. Not much interactional work seems to be expected on his part.

76　Business and Service Telephone Conversations

Example 16: Music school, Frankfurt

01	C:	((telephone rings))	
02	R:	hier hochschule für musik und darstellende kunst guten morgen?	R: this (is) the college of music and performing arts good morning?
03	C:	guten morgen hier ist gerhard dietrich ähm ich wollte fragen ob die frau nase schon im haus ist	C: good morning this is gerhard dietrich e:m i wanted to ask if mrs nase has already arrived
04	R:	.hh frau na:se e: kontrabass	R: .hh mrs na:se e: double bass
05	C:	ja genau	C: yes exactly
06	R:	nein noch nicht	R: no not yet
07	C:	ie- die ist immer im acht vier hundert drei	C: ie- is she always at eight four double oh three
08	R:	ja: aber sie ist noch nicht gekommen	R: ye:s but she hasn't arrived yet
09	C:	ist noch nicht gekommen	C: she hasn't come yet
10	R:	ne	R: no
11		(0.5)	(0.5)
12	C:	ja gut dann. (.) denn kann ich kann ich noten hinterlegen an der portier [(falls)	C: yes good then. (.) then can i can i leave notes at the reception [(in case)
13	R:	[ja selbstverständlich können sie [das machen	R: 　　　　　[yes sure you can [do that
14	C:	[okey ich bin gleich da.	C: 　　　　　[okay i'll be there in a minute.
15	R:	okey dan[ke	R: okay [thanks
16	C:	[ja bis [dann	C: 　　　　[yes see you later
17	R:	[tschü[:ss	R: by:[:e
18	C:	[tschüss	C: 　　　　[bye

(ATD 15 *FFM22M *FFM SCHULE VON MUSIK)

The confirmation check is done this time by a repetition followed by the categorisation of the person searched for (*frau na:se e: contrabass*, line 04). After having received confirmation, R gives the response with the minimum requested information: the person has not arrived yet (line 06). The conversation does not stop here, although the information has been delivered and in this case R cannot do much more. It's C that starts asking other questions: first of all confirmation of where he can find the lady he is looking for. And R, besides confirming the place, repeats the previous response: the lady has not arrived yet, and it looks like he feels his speaker has not understood what he said in the previous turn. This time C echoes by repeating the news and finally R confirms just by saying *no* (lines 07–10). At this point, before C retakes the floor, a gap of half a second occurs and then C reconfirms receiving the news by saying *yes, good then*, but also keeps the channel open and asks for more information (*dann kann ich kann ich noten hinterlegen an der portier (falls...), then. (.) then can I can I leave notes at the reception (in case...)*, line 12). Finally he (C) gets a positive response and he can proceed to the closing part of the call, by saying he is approaching the school.

R's repeating the target request and thus asking for confirmation is a strategy used by R to prepare the response: in the second example this is done because of the particular kind of request made, in the first and third instances (examples 14 and 16) they are also well-suited devices to fill the gaps that otherwise would have occurred if R had just looked for the information requested, the opening hours of the museum, and if the lady searched for had already arrived at the school.

5.2 Request for details

5.2.1 Insertion sequence coming with a request for repetition

The request for a type of service may come in different formats, very simple, or more complex ones, mitigated by pre-sequences and story prefaces. The more complex formats take longer conversational space and may thus be more difficult for R to listen to carefully and store all the important details at once. R of the call below (example 17) seems not to have paid full attention to C's speech, so that he has forgotten the details useful to give a response to the request in the relevant turn (line 04). The call is directed to a car accessories shop in Portsmouth.

Example 17: Car accessories shop, Portsmouth (UK)

```
01   C:    ((telephone rings))
02   R:    >(hallo) car engines?<
03   C:    .hhh eh good(.)afternoon i've-v-got a smashed window on
           a citroen bx .hh >i wonder if i could just buy< a piece of glass to
           p:ut in it (ready) made?
```

```
04   R:    e:hm yeah. what type of bx is it? sorry i forgot?
05   C:    e:::m   nineteen   o:   heitch   reg   can't   remember
           [heitch reg
06   R:    [(°      °) can you tell me what side is it?
07   C:    a: driver's side.
08   (9.0)
09   R:    it's not what i keep on the shelf unfortunately?
10   C:    a okay
11   (1.2)
12   R:    okay i could a: arrange for [sort of=
13   C:                                [.hhhh
14   R:    =tomorrow? afternoon?
15   C:    how much would it be do you know?
16   R:    e:: i would say about eight pounds
17   C:    eight pounds. .hh let me think about it i might just
           go to a glass shop and get them to cut a bit of glass to stick
           in it
18   R:    okay?
19   C:    yeah. okay thanks a lot. [by:e
20   R:                             [°bye°
```

(GDGB 13 *POR50M1 *POR CAR ACCESSORIES SHOP)

In line 04 R initially displays readiness to respond by the response token *yeah*; then he asks about the type of car C is talking about, apologises and gives an account for doing it right after this request for detail: he has forgotten what car they are talking about. On the one hand, the apology may actually signal a repair to a lack of attention, as displayed by R's utterance, by which he is no longer sure whether C has also given the car details in his long turn of talk. On the other hand the apology may just be working as a politeness formula asking for one more detail about the car, i.e. its model, which C has not mentioned in his request. R's request for details *what type of bx is it?* is now misunderstood and responded to by C giving information about the date and registration of the car. This first request is followed by another one in line 06, where this time R narrows down the request by asking which side the broken window is, before being able to give a positive or negative response about the availability of the glass. The response then, that the glass is not available, comes in line 09, after a nine-second pause, during which R does online research in the catalogue of in-store products. C reacts to the news just with *okay*, and leaves the floor to his interlocutor, who retakes it after a 1.2-second gap to offer a solution to the problem. C's *okay* and the following silence are interpreted as a further request for help, made relevant by R's initial *okay* before offering to obtain the glass in a few days' time (lines 12–14). On the one hand, the way in which C says *okay* suggests the relevance of providing immediate feedback to the information provided, but R's not immediate follow-up with more talk suggests that C is left with an unsolved problem and is waiting for a possible solution.

Here the response takes the shape of one of the formats analysed in Chapter 4, in which an increment repairs the lack of service provided in the initial response. The increment, as others in the corpus, comes in a turn unit separate from the initial response, after a receipt token on C's part, and is produced as syntactically independent from the rest of the response. This grammatical configuration of the turn increments shows that when the response is conveyed in two parts, these are also separately and consecutively projected by the speaker. This means that the expansion of the second-pair part is made relevant and projected by R only with receipt of no more talk from C after the acknowledgement token.

But the call is not ended yet, and at this point it is C that opens a short insertion sequence to return to the initial inquiry asking for the price of the glass, before definitively rejecting the offer in line 17: *let me think about it i might just go to a glass shop and get them to cut a bit of glass to stick in it* and thus passing to the closing sequence of the conversation.

5.2.2 Insertion sequence completing the request

Sometimes C's request is formulated with insufficient information for R to allow her/him to provide an adequate response. R then targets the missing information s/he needs to accomplish her/his job. Asking for more details of the request just made is one of the most common phenomena in the corpus. Nevertheless, there are also cases in which the initiation of a short insertion sequence seems mostly to be related to an established conversational ritual, so that, even if not entirely necessary, the request for details is initiated. In the call below (example 18), C's request is apparently complete and sufficiently specific: it is a woman speaking and calling a hairdresser to know the prices for a cut. R responds to the request with another question, not taking it for granted that, if it is a woman speaking and if she does not specify anything else, the information requested should be for her. She (R) thus asks for a specification, *für eine dame?* (*for a lady?*), implying there are different prices for men and women and that the hairdresser deals with both of them.

Example 18: Hairdresser, Frankfurt

01	C:	((telephone rings))	
02	R:	haarstudio franz ritter guten tag?	R: franz ritter hair studio good morning?
03	C:	e: guten tag ich hätte gerne gewusst wie teur ein haarschnitt bei ihnen ist	C: e: good morning i'd like to know how much do you charge for a cut
04	R:	für eine dame?	R: for a lady?
05	C:	>ja für mich<	C: >yes for me<
06	R:	ja dreiundvierzig euro	R: yes forty-three euro
07		(0.2)	(0.2)
08	C:	wie viel bitte	C: sorry, how much
09	R:	dreiundvierzig euro	R: forty-three euro
10	C:	a: achso .h e::m muss ich eigentlich ein termin nehmen [oder ( )	C: a: so .hh er:m do I actually have to make an appointment, [or ( )
11	R:	[ja ja ne	R: [yes yes no
12	C:	e:::m kann ich eventuell später nochmal anrufen dann schaue ich [genau wenn ich genau:	C: e:::m can I possibly call you back again later so that I see [exactly when I exactly:
13	R:	[ja	R: [yes
14	C:	= zeit habe?	C: = have time?
15	R:	mh mh gut	R: mh mh good
16	C:	okey danke [schön bis- tschüss	C: okay thank [you very much see- bye bye
17	R:	[tschüss	R: [bye

(ATD 18 *FFM23F *FFM FRISEUR)

Here the insertion sequence is seen incrementing the initial request and completes it with more detail that is important for R to know and specify. This kind of sequence initiation displays what has been pointed out at the beginning of the chapter, i.e. that moment by moment speakers need to adjust the information they tell their interlocutor.

The second part of the adjacency pair request/response comes after this sequence in a simple format: *ja dreiundvierzig euro (yes forty-three euro*) in line 06. C asks for this to be repeated in the following turn after the gap because of the hairdryer noises in the background that disturb the line, and the conversation carries on later with some more questioning on the part of C asking for different information. Similarly to example 16, the initiation of the insertion sequence does not introduce further complex talk, but is just a means of adjusting the information provided in the request according to the criteria of the service, and the response can then be formulated in simple as well as in more complex ways.

5.2.3 Targeting the request

Similar to the excerpt in the section above, R can initiate an insertion sequence starting a negotiation process with C to better meet her/his request that can occupy longer sequences. In the call below (example 19), the response is negotiated over the whole duration of the conversation. As with some other calls in the corpus, the purpose is to make an appointment at the dentist's.

Example 19: Dentist, Sardinia (Italy)

01	C:	((telephone rings))	
02	R:	>studio mirella buongiorno<	*R: >studio mirella good morning<*
03	C:	e: buongiorno sono: la mamma di chiara rietti volevo prendere un appuntamento per cortesia	*C: e: good morning i:'m the mum of chiara rietti i wanted to make an appointment please*
04	R:	cosa deve fare signora	*R: what do you need to do (madam)*
05	C:	no allora è in (0.2) in cura con il dottor mirella perché c'ha:: insomma l'apparecchietto. .hhh l'ha vista l'ultima volta il due agosto a::m ad oschiri (0.2) e quindi poi non le ha dato appuntamento dal giorno quindi volevo sapere quando la voleva vedere	*C: no well she is (0.2) having treatment with doctor mirella because she's go::t well braces. .hhh he last saw her on on the second of august last time a::m in oschiri (0.2) and so after that day he didn't fix her any appointment so i wanted to know when he wants to see her*

06	R:	mattina o pomeriggio	R:	morning or afternoon
07	C:	di pomeriggio perché va a scuola	C:	in the afternoon because she goes to school
08		(1.2)		(1.2)
09	R:	e verso che ora signora? presto, tardi? me lo deve dire lei °perché io,°	R:	e around what time madam? early, late? it's you who have to tell me °because i,°
10	C:	eh? guardi noi alle quattro:: e mezza:: va bene ci- anche alle cinque?	C:	eh? look we at ha::lf past fou::r is fi- also five o'clock alright?
11		(1.2)		(1.2)
12	R:	nella prossima settimana?	R:	next week?
13	C:	non lo so quan[do::-	C:	i don't know whe[::n-
14	R:	[perché questa setti^mana: ci sarebbe giovedì (1.0) però: alle se:i	R:	[because this we^e:k there would be (a spot) on thursday (1.0) bu:t at si:x
15	C:	no no alle sei [fa	C:	no no at six [she's
16	R:	[troppo [tardi?	R:	[too [late?
17	C:	[allenamento. sì sì no. diciamo intorno alle cinque se va bene	C:	[training. yes yes no. let's say around five if it is okay
18	R:	insomma alle cinque alle quattro e mezza, [cinque	R:	so at five at half past four, [five
19	C:	[sì	C:	[yes
20	R:	e e: anche martedì?prossimo?	R:	e e: also next? tuesday?
21	C:	martedì prossimo	C:	next tuesday
22	R:	alle cinque?	R:	at five?
23	C:	va bene	C:	okay
24	R:	a nome?	R:	what name?
25	C:	e: chiara rietti	C:	e: chiara rietti
26		(1.5)		(1.5)
27	R:	ascolti io ho segnato martedì ventiquattro? ore diciassette	R:	(listen) i have written tuesday the twenty-fourth? at five p.m.

28	C:	va bene:	C:	all right
29	R:	se ci sono problemi mi chiama	R:	if you have problems you call me
30	C:	va bene [grazie	C:	okay [thank you
31	R:	[buongiorno si[gnora	R:	[good bye mad[am
32	C:	[bn-giorno:	C:	[goodby:e

(SNI 22 *SS DENTISTA)

This time, before R can give a satisfactory response to the request for an appointment, she needs to initiate an insertion sequence for deciding, together with C, when to make the appointment, depending, first of all, on what kind of visit C needs (*cosa deve fare signora, what do you need to do*, line 04). C completes the pair by telling a story, by giving details and explaining the reason for the requested appointment, rather than opting for a quick response by just saying the reason for the appointment is a check-up.

The next move is now finding a suitable day and time for both speakers. This takes the rest of the conversation as both speakers now need to align and agree on a common date for the appointment. This is a problematic issue on both parts, as on C's end of the line is the mother calling on behalf of her child who goes to school and cannot miss it, plus she is calling from a place that is 60 miles away from where the surgery is. On the other end of the line, the dentist's secretary who is receiving the call has to try and meet C's needs within the constraints of the agenda and the remaining free slots of the dentist. It is R who first suggests the two daytime possibilities (*mattina o pomeriggio, morning or afternoon*) in line 06, right after C's story in response to the request for details. The choice of the time is then partly left to C, as a consequence of C's listing all the child's commitments. The distance between the patient's home and the surgery makes more relevant R's asking C to give a suitable time when she can afterwards fit her in the free slots for appointments in line 09.

The arrangements take various turn units, in which, first, the time of the day and then the date for the appointment are sequentially narrowed down and confirmed. It is R suggesting a possibility each time, adjusting the suggestion according to the commitments displayed by C, which is promptly confirmed or rejected by C together with an account, as in lines 15–17, where a possible appointment is rejected because of C's daughter's training.

At the end of the negotiation over the appointment (line 27) R sums up the result of the negotiation: *ascolti io ho segnato martedì ventiquattro? ore diciassette (listen i have written tuesday the twenty-fourth? at five p.m.)*, as the closing turn unit and second-pair part of the sequence initiated by C in her first turn of the conversation, and a few minutes previously. This response is then incremented in the following turn (line 29) by R's offer of further availability in case of other last-minute needs and changes. This increment is produced, like most of those produced in a turn other than the one in which the response is delivered, as syntactically independent from the response.

5.2.4 Narrowing down the response to a wide request

Sometimes the request made is not general or straightforward, but rather concerns specific and complex items, so that the conversation in these cases takes the form of a question–answer sequence opened between the initial request and the final response. In these instances R starts targeting the object of the initial question through other questions, quite similar to what happened in the call to the dentist just observed.

A typical and straightforward case is a call to a travel agency to book a flight. The request for booking information always presupposes that the details of the request, i.e. the possible departure and return dates, type of connection (direct/indirect), the price, the number of passengers, their names, etc. are going to be asked for by R when appropriate, while s/he looks at the computer and fills in the reservation. But this kind of sequence is also initiated, as mentioned above, when the reason for the call is complicated. In the excerpt below (example 20) R first starts to provide a response, and asks further questions 'on the way' in order to be able to give a more adequate response. The call is made to a travel agency that commonly deals with special prices for the flight connection between Sardinia and mainland Italy.

Example 20: Travel agency, Sardinia (Italy)

01   C: *((telephone rings)) ((waiting music)) (13)*
02   R: >centro viaggi buonasera sono adele?<   R: >centro viaggi good morning i'm adele?<

Insertion Sequence Followed by the Response 85

03	C:	eh buonasera. (.) senta signora io avrei necessità:: di un'informazione per quanto riguarda (.) la continuità territoriale chi è che ha diritto agli sconti?	C: eh good evening. (.) (listen madam) i' need some information about (.)'territorial continuity' who can get the discounts?
04		(.)	(.)
05	R:	i residenti i nati in sardegna che risiedono fuori.	R: resident people, people born in sardinia that live outside it.
06		(0.5)	(0.5)
07	C:	mh mh.	C: mh mh.
08	R:	mi dica com'è l-la: la persona? ha meno di ventisei anni?	R: tell me how is th-the: the person? is s/he under twenty-six years of age?
09		(.)	(.)
10	C:	no no no io poi volevo sapere se per esempio (.) in caso di: una famiglia separata per esempio	C: no no no i then wanted to know if for example (.) in the case o:f a separated family for example
11	R:	sì	R: yes
12	C:	e:: e:m uno dei coniuge è residente qu[i.-=	C: e:: e:m one of the spouses is living he[re.-
13	R:	[ho capito	R: [i see
14	C:	l'altro in un'altra regione della: sard- della:: dell'italia	C: the other in another region of sard- of italy
15	R:	sì	R: yes
16	C:	e: i::l minore anche se si trova in un'altra regione ha diritto allo sconto?	C: an::d the:: underage child, even if he is in another region has he the right to get the discount?
17	R:	ma è in sardegna il minore?	R: but is the minor in sardinia?
18	C:	no.	C: no.
19	R:	no allora no. sono sono i lati::: beh no aspetti il minore? quanti anni ha? se ha meno di ventisei anni sì?	R: no then no. it's it's the si::des well no wait the minor? how old is he? if he is younger than twenty six yes?

20	C:	sì sì ha meno di-		C: yes yes he's younger than-
21	R:	e allora sì.		R: then yes.
22	C:	e allora ha dirit[to com-		C: then does he have the ri[ght any-
23	R:		[se è così sì	R: [if it's like so yes
24	C:	<ho capito>		C: <I see>
25	R:	perché è pe:r gli emigrati? quelli nati in sardegna? per i residenti per i giovani (.)		R: as it is for emigrants? people born in Sardinia? for the residents for young people (.)
26	C:	°sì°		C: °yes°
27	R:	per i senior. cioè più di sessant'anni		R: for seniors. that is more than sixty years of age
28	C:	°sì°		C: °yes°
29	R:	pe:r eh gli studenti universitari al di sotto dei ventisette anni?		R: fo:r university students under twenty seven years of age?
30		(0.2)		(0.2)
31	C:	ahah.		C: ahah.
32	R:	e comunque se è un minore?		R: and anyway if it's a minor?
33		(0.2)		(0.2)
34	C:	ho [capito		R: i [see
35	R:	[è un giovane		C: [it's a young person
36	C:	ho capito .hh e per quanto riguarda (.) i prezzi (.) e:m alghero roma? quanto costa.		R: i see .hh and for what concerns (.) the prices (.) erm alghero to rome? how much does it cost.
37	R:	>alghero roma e roma alghero andata e ritorno tasse incluse ottantasei virgola ventitre euro		R: >alghero to rome and rome to alghero return including taxes (is) eighty six euro and twenty three cents
38		(0.2)		(0.2)
39	C:	e: invece. olbia roma? [...]		C: a:nd olbia to rome instead? [...]

(CVI 19 *NU42F1 *SS viaggi)

Here, after the answerer's first attempt to reply to the request, a gap of half a second occurs (line 06), waiting for C's reaction to the news. This comes right after the gap, with *mh mh*. Jefferson (1985, p. 206), comparing the use of 'yeah' and 'mm hm', notes that the second exhibits 'passive recipiency', proposing that the co-participant is the current speaker and will go on talking. This token may have different functions and here it displays the features Gardner (1997, pp. 137-8) has noted for 'mm hm' as a weak acknowledging token: it is articulated clearly, occurs in a sequentially relevant position, after a gap, which is in a free space, occurs in a topically coherent position in the development of the topic, which is right after a first bit of information has been provided and the other speaker is waiting for some acknowledgement. It also occurs at a point of possible full grammatical and pragmatic completion. The *mh mh* here displays a weak acknowledgement in that the information provided is not satisfactory for C, who has a precise case in mind to ask about, and leaves the floor to her recipient. The specific instance for which C is actually making the request is not made explicit until R initiates the insertion sequence by asking for the details of the person that should get the discount (line 08). Only now does C take the floor to explain her case, through a sequence in which R takes the floor to signal she is listening and C can go on explaining her case (*yes, I see*, lines 11-15).

Before actually getting to the final response, R asks another question trying to make the person C fit the discount conditions (line 17). Even when R starts giving the new response, this changes according to the information R receives from C and the document she is reading, so that at the beginning the response is negative, but it becomes positive right after (lines 19-24). R now expands the response with an account of why it has changed from negative to positive, by listing the kind of people included in the promotion (lines 25-32). Here the expansion comes in a turn unit different from the initial response (lines 19, 21, 23), which is repeated three times. Although occurring in a separate turn, the extension is syntactically tied to the first part of the response, as it represents a subordinate clause and is linked to the initial part of the response by causality. This type of extension, as Ford et al. (2002a) suggest, continues the action of the preceding talk to which it is attached; it thus does not add anything new to what has already been responded to, but, as in the instance under examination, just provides an account of what has already been said.

5.2.5 Repair initiation

The next case is one in which initiation of an insertion sequence contains an initiation of repair. In the next example (21), in which another bookshop in Portsmouth is called, R has some problems in understanding C's request, probably because it is the first time she has heard the name of the Italian author.

Example 21: Bookshop, Portsmouth (UK)

```
01   C: ((telephone rings))
02   R: hello vivian books?
03   C: .hh hello i was wondering do you have a book by an author
     called grazia deledda
04   (1.0)
05   R: sorry, what was the author's name?
06   C: deledda
07   (0.8)
08   R: can you spell it
09   C: yeah d e l e double d a
10   (8.5)
11   R: °i'll check it for you°
12   C: °okay thanks very much°
13   (3.5)
14   R: (we) don't have any:: i don't think we have any:: in stock.
     oh yes we might have one called liars fo-for love.
15   C: okay you don't have one that's called <canne al vento>,
     (0.2) no
16   R: °i'll check that° (.) a: a:: no[we don't
17   C:                                 [no you don't=
18   R: =we could order it for you
19   C: okay no that it's okay. do you have any boo- any other book
     by italian authors? (1.2) do you know these (   )
20   R: well i'd say we have this we have another book by that
     author but that's=
21   C: a okay
22   R: a liars for (   ). e::m in-in translation. in english (   )
23   C: either in english or in italian.
24   R: yeah. we ha-we have got a bit in english
25   C: ah okay
26   [...]
```

(SNGB 09 *IR24F1 *POR BOOK SHOP)

Here the object of the request constitutes the trouble source, and C seems to be well aware of the fact that the author she is looking for may not be known to the British R and the request may thus be problematic. C is in fact asking for a book by a Sardinian author and actually states only half of her request, starting with the author before saying what book she is precisely looking for. R initiates the insertion/repair sequence one second after the request has been made, by first asking for repetition of the author's name, which is followed, after almost another one-second gap after C's repeat, by a request to spell the name (lines 04–08). The trouble is then apparently solved and concluded by putting C on hold, while online research is carried out.

As in other cases, the response comes when R comes back on the phone from the search when the research results start appearing on screen, in line 14. This is why, similar to the call to the travel agency analysed above (example 20), the response changes while it is being uttered. What is interesting here is the way in which the response changes, through the use of certainty verbs first (*we don't have any*), then switching to personal opinion and less sure grounds for stating what is being said (*I don't think*), thus opening the way to the change for the other possibility, that they do store some books of this rarely requested author. The change of state is marked by R's surprise herself, *oh yes*, and is still characterised by the use of a modal verb stating uncertainty (*we might have one*).

At this point C comes in again asking for the book she is looking for (line 15), which she had put on hold trying to first overcome possible trouble with the author's name. This second request already orients to a negative response, which comes after another online research and is then echoed in overlap with R's actual 'verdict'. Now R, reaching completion of the response to the request, extends it with the offer to order the book, a ritual for the British. The extension comes in a second turn of response because of the overlapping talk with C that echoes the negative response. The extension is promptly produced by R and is syntactically independent from the core response. Similar to other syntactically independent increments, this results in an action separate from the core response, that of offering a solution.

5.3 Cross-cultural comparison: actions in the insertion sequence

Summing up the aspects analysed in this chapter, the first to look at is the use of the insertion sequence that delays the actual response.

The initiation of an insertion sequence is restricted to cases in which the request made by C is not clear or detailed enough so that R needs to restrict it by asking some questions targeting its object. Otherwise, speakers try to keep the conversation quick and smooth, and do not take the time to delay it dealing with the issue presented by C. In this case then, in agreement with what was observed for the openings of service encounters (cf. Bowles and Pallotti, 2004; Pallotti and Varcasia, 2008), the time pressure issue seems to significantly influence the way in which the conversation is managed, and restricts the delay of the response only to cases in which this is inevitable.

On the other hand, asking for confirmation is much less used, constituting 21 per cent of the calls with an insertion sequence before the response, as Table 5.1 shows. Confirmation checks, as pointed out above, also seem to be used more often by English Rs, used in 30.4 per cent of the cases and much more rarely by Rs of the other two languages, namely 18.7 per cent of the cases in Italian and 11.1 per cent of the cases in German.

The production of confirmation checks or requests for detail in the insertion sequence imply different things. Delaying the response with one or more detail requests implies, in fact, that the request as it has been produced is complex or provides incomplete information, or needs to comply with some sort of schema, such as in the case of a flight (pre-)booking, etc. In contrast, the production of confirmation checks seems to display a strategy for delaying the response and taking time to find the appropriate one. Such delay can also be understood as wasting precious time, and money, though.

The cross-cultural comparison does not point out any other significant differences among the three languages other than those mentioned above, but rather describes a similar frequency among English, German and Italian calls in the initiation of insertion sequences asking for more

Table 5.1 Actions performed by the insertion sequence (%)

Percentage of all insertion sequences	Details request	Confirmation check
English (N = 23)	65.6	30.4
German (N = 18)	88.9	11.1
Italian (N = 16)	81.3	18.7
Total (N = 57)	79.0	21.0

details about the request. The insertion sequences are similar whether the request is felt as incomplete, but clear, or if it is just complex, or unclear.

5.3.1 Cross-cultural comparison: actions in the response

Another aspect to look at is thus the format the response takes after the insertion sequence. The analysis of the excerpts has shown variability in the production of the responses after the insertion sequence, either a request for more details or just for confirmation. Table 5.2 summarises the kinds of actions initiated to complete the request–response adjacency pair after the first-pair part has been clarified and specified.

The calls in which the second-pair part of the request–response sequence is initiated represent 40.4 per cent of all the calls collected. When the insertion sequence has been concluded, R proceeds to deliver the response, and does this using the same response strategies described in Chapters 3 and 4. The response is delivered in a simple way, which is that R limits her/himself to provide the relevant information requested in 14.2 per cent of the calls. In contrast, the majority of the calls displaying this sequential response format were found to be produced in a more elaborate way, in 26.2 per cent of the calls. This means such calls were provided together with an increment through which the response was complemented by the addition of more information, or justified, or an alternative solution was offered, or complemented by doing two or more of the above actions.

Italian and British Rs tend to increment the response further, 26.8 and 30 per cent respectively, even after the response has been delayed and often pre-specified in the request for details. They do this more

Table 5.2 Actions in the response after the occurrence of an insertion sequence (%)*

Response	Simple response	More info	Alternative solution	Account	Combined actions
English (N = 50)	16.0	10.0	2.0	6.0	12.0
German (N = 42)	21.4	7.2	2.4	11.9	–
Italian (N = 49)	6.2	8.2	2.0	10.2	6.2
Total (N = 141)	14.2	8.4	2.1	9.3	6.4
			26.2		

* The values shown in the table are partial and they add up to those of the tables shown in Chapters 4–6. A table showing all formats of response used and their frequency can be found in Chapter 7.

92 Business and Service Telephone Conversations

often than just provide a minimal response, i.e. 6.2 and 16 per cent respectively in Italian and English. In contrast, Rs in Germany do not seem to follow this trend and apparently increment the response after the initiation of an insertion sequence as often, i.e. in 21.5 per cent of the calls, as when they respond minimally, i.e. in 21.4 per cent of the calls.

But what does R of the three languages do when s/he increments the response after the insertion sequence has been completed? The two actions performed more frequently in all three languages are the increment with an account, an explanation of the response just given (9.3 per cent), and the addition of more information (8.4 per cent). The offer of an alternative solution is rarely found as a strategy of response increment alone (2.1 per cent, one call in each language), but is rather more often produced together with additional information or an explanation of the problems of the product/service requested (6.4 per cent).

5.4 Summary: insertion sequence followed by the response

This response format gives us a more complicated picture than one could expect. If in Chapter 4 the issues that influenced the categorisation of the responses were mainly dependent on their expansion, and the degree to which this was incremented, there are now at least three factors that need to be taken into account. First of all, we have seen that the way in which the request is formulated, whether it is precise or vague, plays a decisive role in the way in which the conversation then develops. A second factor is given by the insertion sequence initiated and its length. It has been noted that there can be calls in which the insertion sequence is just a brief deviation from the course of the conversation, and the second part to the request pair comes soon in the call. There are also other instances, though, in which the insertion sequence actually takes up most of the exchange between the customer and the shopkeeper, which have been pointed out. Thirdly, how the first response is then formulated influences the overall format of the sequence under examination here.

The examples observed point out another issue, which is how the three factors highlighted above interact with each other and are combined in the single cases. We may have difficult requests (including requests about rare items, such as the call asking for the book by the Sardinian author, example 21, or complicated requests for information, such as the request for discounts in example 20 or the one requesting the appointment at the dentist's, example 19), followed by long insertion sequences and then by an expanded response, or we may still have difficult requests, long insertion sequences, but then just simple responses. But we may also find not so complicated requests, short insertion sequences followed by simple responses (see example 18). The combination of the three factors can be extended further, thus providing different degrees of structural complexity. The point here is that all of these responses display a much more complex structure compared to the excerpts in Chapter 3, and they thus need to be considered differently, and they need to be considered when training for service providers' staff and call centre operators is provided. This issue will be dealt with in Chapter 8.

In the previous section, the occurrence of simple and incremented responses has been pointed out, as well as the actions that were initiated by the increment. As in Chapter 4, such expansions can be produced either in the same or in a different turn from the response and can be formulated as grammatically dependent or independent from it. Table 5.3 shows the format of the response after the insertion sequence, distinguishing among simple responses and responses expanded by the different types of increments, free constituents or proper extensions produced in the same or in another turn from the actual response. The percentages showed in parentheses indicate the proportion of calls in which the insertion sequence is a confirmation check.

Table 5.3 Increment types in response format: insertion sequence + response (%)*

Calls in the corpus	Simple response	Free increment		Extension	
		Same turn	Other turn	Same turn	Other turn
English ($N = 50$)	16.0 (6.0)	6.0 (4.0)	18 (4.0)	6.0 (2.0)	0.0
German ($N = 42$)	21.4 (4.8)	2.4	11.9	7.2	0.0
Italian ($N = 49$)	6.2 (4.1)	6.2	16.3 (2.0)	2.0 (2.0)	2.0
Total ($N = 141$)	14.2 (4.9)	4.9 (1.4)	15.7 (2.1)	4.9 (1.4)	0.7

* The values shown in the table are partial and they add up to those of the tables shown in Chapters 4 and 6. A table showing all types of increments used in the cross-cultural corpus and their frequency can be found in Chapter 7.

94 Business and Service Telephone Conversations

Table 5.4 Type of response following the insertion sequence (%)

Response after the insertion sequence	Positive	Negative
English (N = 23)	73.9	26.1
German (N = 18)	77.8	22.2
Italian (N = 16)	62.5	37.5
Total (N = 57)	71.9	28.1

Simple responses follow the insertion sequence more often in English and German than in the Italian calls, accounting for 16 and 21.4 per cent compared to 6.2 per cent in Italian. As already pointed out in the previous section (5.3.1), in the Italian calls first-pair parts responded to after an insertion sequence receive extended responses more often than those formulated just in minimal formats. These are produced as free increments in separate turns, representing 16 per cent of the British calls in the corpus. British and German Rs mostly complete the adjacency pair by expanding their responses through free constituents in separate turns, accounting for 18 and 11.9 per cent respectively. But for the Germans in the corpus the response after the insertion sequence mostly comes in a simple format (21.4 per cent). The news delivery, after the interruption of the normal adjacency pair completion flow in a simple format, seems to signify that enough has already been done by R, and that probably most of the other issues concerning the object of the request have already been answered within the insertion sequence. Another reason why the insertion sequence is often followed by a simple response may be the fact that most of the calls occurring with this format receive a positive response. This means that usually it is only calls in which the service requested cannot be provided that get an increment to the response proper, as in example 17 above, in which there is an offer of a solution. Table 5.4 shows the frequency of different response types following the insertion sequence.

Table 5.4 shows that 72 per cent of the requests that have been responded to through the initiation of complex sequences receive a positive response, and only 28 per cent receive a negative one. This indicates that the specification of the initial request C has made often leads to a positive outcome, and the initial complexity is resolved precisely in the deployment of the insertion sequence.

The deployment of the request–response sequence following the structure presented in this chapter represents the largest part of the corpus, meaning that after all what matters first is that both speakers understand each other and speak about the same thing. There is still the remaining part of the calls in the corpus to be analysed, those representing a different deployment of the request–response sequence, sometimes arriving at the opposite of what was observed in the calls in this chapter: it is C leading the conversation throughout. This kind of call will be considered in the next chapter.

6
The Caller Leads the Conversation

This chapter, as the preceding one, combines some of the features analysed before, such as the simple, the extended and the delayed responses together with new features. Here the emphasis is put on a specific conversational phenomenon that characterises the missing part of the data in the corpus to be analysed. As mentioned in Chapter 3, there are calls that receive a response to the request in a simple format, but then C, instead of aligning and moving towards the closing of the conversation as in the excerpts analysed there, prompts a continuation of the exchange. The result often is that C leads most of the conversation and s/he produces more talk.

6.1 The caller soliciting more talk from the receiver

The formats analysed in the previous chapters include all the possible sequential deployments that depend on R's initiation of some action in response to the request made by C, an extension to the response or an insertion sequence preceding the response. These represent 85.1 per cent of the corpus, which is the majority of the development of the calls after the opening is almost entirely left to R's initiation. The missing 15 per cent of the calls are characterised by a clear-cut phenomenon. In these calls the request receives a very simple response, similar to what happens in the calls in Chapter 3, but the original request–response adjacency pair is not accepted by C as satisfactorily completing the business of the call and the exchange does not proceed to the closing sequence as it does in the calls observed before (Chapter 3). The following example (22) is one of the most representative of R's options in this category. The call is made to a hairdresser in Sassari (Italy). The hairdressing salon is not situated close to the university, although other places in the same area give special discounts for students anyway.

Example 22: Hairdresser, Sardinia (Italy)

01	C:	((telephone rings))	
02	R:	>alma buonasera<	R: alma good evening
03	C:	buonasera scusi un'informazione >voi siete: mh fate sconti pe:r studenti< per tagli [e piega	C: good evening excuse me some information >are you: mh do you do discounts fo:r students< for cuts [and hair set
04	R:	[no hh	R: [no hh
05	C:	niente. non siete neanche affiliati arco per caso?	C: nothing. are you not even affiliated to arco by any chance?
06	R:	no no	R: no no
07	C:	ah ho capito. quanto costa comunque [per il-	C: ah i see. how much is it anyway [for a-
08	R:	[ventidue euro	R: [twen-ty two euros
09	C:	ah? va bene.	C: ah? that's okay.
10	R:	( )	R: ( )
11	C:	a(.)scolti allora richiamo per un appuntamento	C: (listen) then i'll call you back to make an appointment
12	R:	va [bene	R: that's [okay
13	C:	[va bene? grazie ar[rive-derci	C: [is that all right? thanks ar[rivederci
14	R:	[arri-vederci arriv°ederci°	R: [arriv°ederci° arrivederci

(ATI 03 *SS23F3 *SS PARRUCCHIERE 3)

Here, in fact, the initial request is made by C in line 03 as a multi-unit turn: it is prepared by a first TCU component, a request preface (*scusi un'informazione, excuse me some information*, line 03) and then followed by the actual request reformulated twice with a change of verb (*voi siete mh fate sconti ..., are you mh do you do discounts ...*). The adjacency pair does not need an elaborate response, being a yes/no question, but could also come as a complete sentence (no we don't have any sort of discounts). Here R answers minimally by just saying *no*, and does not seem to be feeling the need to explain why they do

98 Business and Service Telephone Conversations

not have discounts or add anything to her reply. In contrast to the calls observed in Chapter 3, here C retakes the floor and tries to keep the channel open by first reformulating the response (*niente, nothing*) as a sign of receiving the news and then reopening the conversation through another question, related to the first, which is by asking whether she could get concessions through other means and associations (line 05).

After this second reformulation of the request and receiving a second minimal but clear response through a double repetition of the negative response (line 06), C finally gives up asking for more concessions and displays the upshot of the preceding action by showing her understanding (*ah ho capito, ah I see*, line 07), the reception of the negative news and impossibility of getting any sort of discounts also highlighted by the elliptical response R has provided. Right after this C switches to another topic and asks for the price she would then have to pay. Again there is a minimal response from R, and s/he still does not seem to make much effort to cooperate with the ongoing talk (lines 07–08). R's minimal reaction to all C's prompts may be explained by the conversation occurring at a busy moment, so that she seems to be looking forward to the conclusion of the conversation as soon as possible, without giving much space for other talk.

From C's point of view the expectations seem to be different. When R does not extend her/his turn by giving more information or apologising for not offering the service requested, it can be C that keeps the channel open and solicits an extension of the response by asking for more details, as C does in this example. C takes three more turns after the initial request before finally taking leave and moving to the closing of the call. She finally starts orienting to the closing in line 11 by promising a further call, and by asking confirmation that this is fine with R in the following turn. By giving minimal responses and not doing anything to mitigate the information she is giving, R is 'pointing' to the upshot of the call. Nonetheless she cannot herself officially initiate the closing section as she has to wait for C to ask all she needs and orient to the closing as well.

Both speakers seem to orient to a shared conversational rule, typical of the conversation on the phone (see Schegloff and Sacks, 1973, p. 310): that is, the person who is calling has the right to access and initiate the closing sequence. Here R seems to follow such a rule more tightly by her waiting for the other speaker to orient to the closing and open this sequence, despite her minimal way of responding to all of C's solicitations for more talk and the current activity interrupted by the call. In this context C's only right to access the closing sequence is also a premise, as a customer, to her/his weighting and deciding that the information received in response to his/her inquiry was satisfactory and the conversation can proceed to the closing sequence. From a 'business' point of view, in this case R's attitude could be considered as failing the service offer, and doing nothing to acquire a new customer, though respecting the main conversational and politeness rules. The difference in comparison with the sequences analysed in the previous chapters seems that R, instead of trying to be 'nice' to the possible customer, seems to care more about the job that has been interrupted by the call.

The conversational structure of this call follows the scheme of question–answer series, namely here Q1–A1, Q2–A2, Q3–A3, and so on, in which the question is usually extensive, and from the second question onwards it is prefaced by an acknowledgement of the previous response, whereas the answer is rather elliptical and minimal.

Not all excerpts belonging to this category display such ellipsis in R's responses, but sometimes R's reaction and spontaneous take-up of the conversational topic are slow, and need to be solicited by C's questioning anyway. This seems to work in the direction of a gradual growth of the conversational exchange. The next excerpt is a German call again to a hairdresser, so the requests made are quite similar to those in the previous example (22). As mentioned in Chapter 2, it is quite common in Germany to have apprentices working in hairdressing salons.

Example 23: Hairdresser, Cologne

01	C:	((telephone rings))	
02	R:	>hier friseur salon scheler patrizia am apparat?<	R: hairdresser studio scheler patrizia speaking
03	C:	guten tag ähm ich wollte nachfragen ob sie modelle nehmen für dauerwelle	C: good morning ähm i wanted to ask you if you take models for perms
04	R:	ja >nehmen wir<	R: yes >we do<
05	C:	ah. und wann kann man? also an welchen tagen ist das?	C: ah. and when is it possible? so which days is it?
06	R:	.hh uhm. das wurde gehen jeden zweiten dienstag und jeder zweiten donnerstag	R: .hh uhm it would be every second tuesday and every second thursday

07	C:	und wie teuer ist das denn	C: and how much does it cost then
08		(1.5)	(1.5)
09	R:	u::m (1.2) dauerwelle normal .hh das (reinisch berate) ehm würde sie zweiundsiebzig kosten und ä:m die modelle bezahlen bei uns die hälfte also >wäre das denn<	R: e::m (1.2) the regular/ normal perm .hh it ( ) ehm that would be seventy two and ä:m the models pay here half the cost it >would be then<
10	C:	einunddreißig	C: thirty-one
11		(1.5)	(1.5)
12	R:	ja sechs und dreißig wäre das denn	R: yes it would be thirty-six then
13	C:	eh: ah ja. (.) ehm (.) und dann muss man sich bei ihnen ein termin holen.	C: eh: ah yes (.) and then do i need to make an appointment
14	R:	.h genau	R: .h exactly
15	C:	mm .h ah ja gut. vielen dank	C: mm .h ah yes good. thank you very much
16	R:	bitte schön	R: you're welcome
17	C:	wieder hören	C: good bye
18	R:	wieder hören	R: good bye

(CVD 12 *K22F *K Friseur1)

Here the structure of the conversation follows the pattern outlined above, of a chain of question–answer sequences, but with a few differences with respect to the first example presented. Firstly, if we look at the first of those Q–A sequences, the response to the first request, here again a yes/no question, comes in a minimal format, but not an elliptical one such as in the call above (example 22). The 'yes' particle is accompanied and completed by the verb, thus forming a complete sentence.

C's reaction to the response, with *ah*, is just a variation of the change of state token analysed by Heritage (1984a), and is thus displaying a change of state in the expected response. Here *ah* works like the 'oh' described by Heritage (1984a):

> a means by which recipients align themselves to, and confirm, a prior turn's proposal to have been informative. Furthermore, by the

addition of specific types of turn components, such as assessments or requests, recipients can proceed to treat the local trajectory of the informing as complete (with assessments) or incomplete (with requests for further information). (Heritage, 1984a, pp. 304–5)

The request following the *ah* token is thus treating the information received as not yet complete. The second request is then first formulated generally and immediately after being narrowed down to a specific question (*und wann kann man? an welchen tagen ist das?, and when is it possible? so which days is it?*, line 05). Here R again replies with a complete sentence, by giving the details of the days of the week they accept models.

The third question, the one about the price, now receives a delayed response, as it needs to be checked in the price list. R thus formulates her response to this question (line 09) with a long preface in which she explains what sort of operation she is doing to calculate the price for models. This preface, as in other calls, allows R to take the necessary time to calculate the actual price C needs, avoiding a longer gap. C, on her part, participates actively in the cognitive operation that is going on and tries to complete herself the calculation and her speaker's turn. Here the participants' collaboration in the construction of the turn is primarily important for the completion of the action initiated by R in the calculation of the price, but is also relevant in completing the syntax of the sentence under construction, by assigning an immediate object to the suspended predicate.

The confirmation and correction of the price suggested by C come in line 12 after 1.5-second gap during which R finally calculates the precise price in silence. C then acknowledges the information given through another change of state token, followed this time in the first instance by another acknowledgement token *ja* (*yes*). The production of such tokens in this position is closely associated with the acceptance of the new counting provided by R by her interlocutor as a correction. Both speakers then proceed to the next and last question which is this time responded to minimally with just one word, *genau* (*exactly*).

This excerpt differs slightly from the first one observed in that R reacts to the solicitations. Here in fact R's responses, though still tightly constrained to the object of the request made, are formulated simply but not elliptically. Moreover, this time R seems to increase her participation in the conversation as long as the call goes on. She, so to speak, reaches the higher point of her conversational involvement while responding to C's third question. However, one of the reasons why she (R) actually produces a long preface before giving the price asked for is in order not to leave a longer gap in the exchange, before the relevant second part has been delivered. In this way she is also accomplishing a complementary action by giving more information and details about how the price is calculated for the service requested.

The preface to the third answer (line 09) gives a more complex structure to the sequence. This is also a feature that accounts for dispreference (Pomerantz, 1984), whereas on the whole the preceding Q-A sequences display simple characters and thus features of preference.

The third example to be analysed is taken from the British calls, and though there are only three excerpts in English displaying this format, they all display the same structure that moves one step further from those just analysed. The call is made to a florist in Lancaster that deals with flower delivery with international chains.

Example 24: Florist, Lancaster (UK)

```
01   C: ((telephone rings))
02   R: good afternoon? charter square florists?
03   C: .hh oh hello i was just wondering? do you do ahm deliveries?
04   R: we do
05   C: octo? okay great. is that part of interflora chain or are
     you separate to that?
06   R: we- we're a teleflora (street) chain
07   C: oh: i see okay ehm .hh can you give me just some idea of
     the sort of price range eh just a: sort of mh: normal general
     purpose bouquet
08   R: where are you wanting to send [it to
09 . C:                                [.hh be be local be
     within lancaster
10   R: okay well a: we start at eighteen?
11   C: mm hm
12   R: and there is two pound fifty
     delivery char[ge depending where it is
13   C:           [right
14   R: if it's [further out
15   C:         [yeah
16   R: then it gets to be a little bit more [(deeply)
17   C:                                      [yeah?
     o::kay okay that's great that gives me a general idea. then right
     i'll get back to you when i've decided : e::m exactly what i want
     then ((slightly laughing))
```

18   R: okay?
19   C: okay then thanks for your help? [bye
20   R: [by:e

(CVGB 23 *LON33F1 *LCS FLORIST)

This example seems to show another aspect of why C seems to lead the whole conversation through a series of questions. Here the conversation seems to follow a pattern that is built up step by step, or in other words, which proceeds to the next step after each response to the questions posed by C. Following the exchange question by question, as in the above excerpts, the first one is a yes/no question and receives a simple response (line 04, *we do*). Here the information provided is regularly acknowledged through the token *okay*, used in all instances. The second request comes as a consequence to the first response, as in the example above (23), showing that the information conveyed is not yet sufficient, and through another yes/no question type C asks for a specification about the kind of deliveries. And the response to this second request also comes in a minimal and simple format (line 06). The Q–A sequence series is interrupted in the next request that opens up a larger conversational project inquiring about price ranges. The actual response to the question asked in line 07 by C comes in line 10 and is finally completed by line 16. Before that line, the adjacency pair embeds an insertion sequence initiated by R asking about the details of the delivery. As soon as the embedded insertion sequence is closed, R provides the second-pair part to line 07 together with an extension in the following lines. The relevant information comes first by R saying what the minimum price is for making a delivery. The response is then incremented by R saying how much the charge for the delivery is and by giving an account of why and when this would be made. The extension is uttered in a turn different from the initial response, during which C displays participation and reception in overlap with R's talk and it comes as syntactically independent from the initial response.

Only C's first token, the continuer *mm hm* in line 11, comes at a point of possible completion of R's talk, which is uttered with final rising pitch contour and thus seems to seek back-channelling. By intervening with such receipt tokens (*right*, line 13 and *yeah*, line 15) through the further continuation of the response, C seems to display satisfaction with the information received at that point and an initial orientation to the closing of the conversation. Such orientation is finally made explicit

104 Business and Service Telephone Conversations

in line 17, when C proceeds to acknowledging the information received and then to closing the exchange.

This excerpt is slightly different from the first two in the development of R's reaction to C's solicitations and further questions. As noted above, neither of the Rs in the Italian and German calls extends their response to all solicitations prompted by C, in contrast to what the British R does here by producing a complex project at the third request through the embedded insertion sequence and an expansion of the displaced response. It is important to note that the kind of request that has been incremented is not different from those made in the previous two examples and is once again asking about price.

6.2 Cross-cultural comparison: caller leads

A first observation while looking at the corpus cross-culturally is that the proportion of both German and Italian calls displaying this sequential deployment of the request–response is higher than that for British English; they represent about two-thirds more each. There are ten calls in German and eight in Italian displaying this format compared to only three in English. The results of the analysis of this category of response could well be confirmed by a further analysis on a larger sample of data. Such a larger investigation might also look into whether this format is typical of the interactions with British florists (all calls in this format are made to florists) or whether it is just a pattern followed in general by service encounters in response to subsequent requesting by C, as it seems to be for the German and Italian data.

All the British conversations, as the one just observed (example 24), follow the same pattern in which R changes the Q–A sequence format by initiating new actions in response to the third request. The conversation then takes a format of this kind: Q1–A1, Q2–A2, Q3–Q4–A4–A3, etc.

The German calls in this group follow a trend similar to the British ones, as in six calls out of nine R initiates and expands her/his response after C's third solicitation. In contrast, Italians seem to go for either of the possible choices: they do not extend and do not initiate new actions in response to C's initiations, and give simple responses to all prompts in four calls out of eight. And in the remaining four instances the Italian R initiates new sequences, but does not seem to be systematic about the time for doing this yet, as in the data at hand s/he does this at different moments, not limited to the third question, but in moments before this as well. Table 6.1 summarises the distribution of the expansions at

Table 6.1 R responses to C's solicitation of more talk (%)

Response format	Simple	Second position expansion	Third position expansion	Other position expansion
English (N = 3)	0.0	0.0	100	0.0
German (N = 9)	33.3	0.0	66.7	0.0
Italian (N = 8)	50.0	12.5	12.5	25.0

Table 6.2 Actions in the delayed response expansion (%)*

Response	Simple	Apology	More info	Account	Combined actions	Total
English (N = 50)	–	–	6.0	–	–	6.0
German (N = 42)	9.5	–	4.8	4.8	4.8	23.9
Italian (N = 49)	6.1	2.0	6.1	2.0	0.0	16.2
		0.7	3.6	2.2	3.5	
Total (N = 141)	4.9		10.0			14.9

* The values shown in the table are partial and they add up to those of the tables shown in Chapters 4–6. A table showing all formats of response used and their frequency can be found in Chapter 7.

different times in the conversation. Neither in Italian nor in German is this variation statistically significant.

Another point is represented by how Rs actually expand the responses. The next table (Table 6.2) summarises the kinds of actions initiated by Rs to expand their response, when they do expand it. Interestingly, the offer of an alternative solution was suggested in only one call in German, and only together with a previous description of the problem the request raised. Otherwise Rs in all three languages, when more talk was solicited and responded to with an actual increment of their conversational contribution, did this more often by adding more information about the service requested.

British Rs who incremented their responses always did this by beginning with a question in response to C's question, and the response was provided together with the addition of more information. Italian and German Rs provided, as the only resource, more information as an expansion for the calls with a delayed increment, respectively in 6.1 and 4.8 per cent of the calls. Germans also offered an account or combined more than one action together in 9.6 per cent of the calls, evenly distributed (4.8 and 4.8 per cent).

The table also confirms the trend in the way Rs respond to requests in service encounters: they tend, in general, to increment their talk, even when they are solicited by C. The distribution shown provides an overview of the spread of choices Rs made in the calls analysed when facing conversations in which their interlocutor prompted more talk in receipt of their quick response.

6.3 Summary of features: caller leads

The calls observed here point out, through a different sequential development of the request–response sequence, the expectations of the people calling, which is the need to have a conversation that is long enough to display not just interest in the exchange of goods and money, but also shows interest and attention to the other person's needs. R's lack of production of more talk and immediate extension of the response is here complemented and mitigated by C's solicitations and follow-up questions about the same object of the initial request. C does follow up the first request also after the response has already been expanded, thus asking for something else, but these instances were not taken into consideration. The calls analysed here display a specific feature, that is C solicits more talk in reaction to a response formulated minimally. Differently from what happens in the calls observed in Chapter 3, here the participants do not orient towards the closing sequence straight away, but they keep the conversation open, and strikingly this time the person who orients the conversation to the closing is C, not R. C's solicitations and initiation of requests following the first one are for one reason: R's initial response is treated as not yet sufficient and C's asking for more needs to be considered still as part of the same action initiated by the first request, thus resulting in the display of a more complex structural organisation of the sequence. The cause of the display of a more complex structure is here found in C's insistence on getting more talk with regard to the request made to R. When C succeeds in this aim, and R increments the response, s/he can do this through the initiation of an insertion sequence after the first and/or second response has already been delivered and then incrementing the response by using the resources outlined in the previous chapters:

- Adding more information;
- Providing an account;
- Offering an alternative solution;
- Initiating an insertion sequence through which the request is specified; and combining one or more of these actions.

Table 6.3 Types of increments in solicited expansions of the response (%)*

Calls in the corpus	No expansion	Free increment		Extension		Total
		Same turn	Other turn	Same turn	Other turn	
English (N = 50)	–	2.0	4.0	–	–	6.0
German (N = 42)	9.6	7.1	7.1	–	–	23.8
Italian (N = 49)	6.1	2.0	8.2	–	–	16.3
Total (N = 141)	4.9	3.5	6.5	–	–	14.9 (N = 21)

* The values shown in the table are partial and they add up to those of the tables shown in Chapters 4–6. A table showing all types of increments used in the cross-cultural corpus and their frequency can be found in Chapter 7.

As in the previous chapters, the response is incremented either through constituents syntactically tied to the first part or independent from it. Table 6.3 shows the use of free increments and extensions in the calls analysed in this chapter.

The increment of the response in such a sequential position, which is solicited by C, was produced only as syntactically independent from the core response, either in the same turn, in 3.5 per cent of the calls in the three languages, or more often in another turn, in accounting for 6.5 per cent of the calls here. None of the calls expanded was produced through an extension proper, and another 4.9 per cent were not expanded at all. The grammatical configuration of these calls suggests that the increment of the response is not just prompted by the solicitation of C, but also needs to be thought separately from the initial response. This was conceived as minimal and quick by R from its very beginning for different reasons, and first of all because of the activity the conversation interrupts (Hopper, 1991).

C's solicitations of more talk on the part of R suggest an interpretation of the speakers' expectations when making a service encounter on the phone. They are that both parties will try to make the encounter highly focused on the speakers' needs and expectations, and avoid abruptness, despite the implications of 'time is money'. While in the previous chapters it was R who oriented to an expansion of the request–response adjacency pair, here it is C orienting to expansion, and succeeding most of the time.

7
The Different Response Formats at One Glance

This chapter will put together the results described in the analytical chapters (3-6). It will summarise the main features in the analysis of one sort of interactional sequence, i.e. the request-response adjacency pair. Results from the quantitative analysis will also be put together, showing the various formats of response produced by the Rs in the corpus. Finally the chapter will also draw up the results on the analysis of the grammatical configurations of these turns.

7.1 Features summary

The overall analysis of the request-response sequence analysed in the previous chapters has followed the criteria of structural complexity. What follows is a list of features accounting for sequence construction in the request-response adjacency pair in telephone service encounters, starting from the simplest ones (cf. Chapter 3):

- Production of direct answers to the question asked by C in the form of either minimal and elliptical response tokens (*yes, no*) or a full sentence (*yes we do/have/ stock it, no we don't*).
- Responses to requests for orders through the production, once again, of response tokens.
- Turns of talk briefly emphasised through immediate repetition of the response.

Such features were observed to account for a minimal part of the calls analysed in the three languages, whereas the rest of the conversations were found to display more complex formats distinguished by the way in which they were produced and vary in complexity. The features that come into play are therefore turn expansions and delays through the initiation of an insertion sequence. The increment of the response turn, as described in Chapter 4, may be made through the addition of more information about the service requested; by giving an account of, or by offering an alternative solution to, the product searched for, or by a combination of these strategies, e.g. the offer of an account that justifies and explains the response provided, then followed by the offer of an alternative solution. Aston (1988a), Zorzi (1990) and Brodine (1991), working on bookshop service encounters for the Pixi project, also found the offer of an alternative solution as part of the remedial work initiated by the shop assistant when the book searched for was not available and the customer's request could not be met.

The new feature, which was found in the data and which needs to be added to the ones just listed, is the addition of more information about the product requested, often in requests that are being satisfied. These actions were found to be produced as single-response expansion strategies but also as a combination of actions. The following list summarises the main response formats analysed:

- Responses produced immediately after the request, i.e. at a relevant place in the interaction, but formulated together with an expansion in which new information, or a justification, or an alternative solution were added to the response and sometimes also accompanied by an apology when the request could not be satisfied.
- Responses delayed by an insertion sequence specifying the object of the request and then responded to in a minimal format, with simple response tokens and full sentences constraining the object of the request.
- Responses delayed by the initiation of an insertion sequence and then produced with an increment of the type described in the first point.
- Responses expanded only after solicitation by C through further questioning after an initial simple response.

Responses were occasionally delayed by the occurrence of gaps between turns and/or the production of fillers, which have been described as being an index of dispreferred features (cf. Levinson, 1983; Pomerantz, 1984; Schegloff, 2007). As shown in the above chapters, in the data the delay of the response by the use of fillers, or the occurrence of gaps, is not systematically associated with the preparation of dispreferred responses and do not prompt disaffiliation. More often these gaps are due to the unavailability of the information requested. Another issue concerns the use of apologies, found to be mostly used by British Rs and sometimes by Italians to accompany the delivery of so-called 'bad news'.

7.2 Overall description of all formats of response

Table 7.1 shows a summary and the respective frequency in percentages of the different formats of response observed in the corpus overall, making no distinction across the languages of the interaction.

The table shows the occurrence of structurally simple responses in 14.2 per cent of the calls in the corpus, whereas the remaining 85.8 per cent of conversations display a more complex structure. The participants in the telephone service encounters in Great Britain, Germany and Italy use structurally complex actions more often than simple ones. Among the more complex structures of the responses, the table includes the different formats described in the previous chapters (3–6). Such a categorisation was made by choosing among the different strategies of response expansion analysed. Among these strategies speakers more often chose to provide an account of the delivered news, either alone in 22.9 per cent of the calls, or combining the account together with another action, such as often the offer of a solution, in another 14.8 per cent of the conversations. When the response is extended through the offer of an alternative solution, this was used as the only action in 9.9 per cent of the calls, of which 2.1 per cent was previously introduced by the initiation of an insertion sequence, whereas more

Table 7.1 Actions in the response: the corpus overall (%)

Response format	Total ($N = 141$)	
Simple response		14.2
Insertion sequence + simple response		14.2
Response + apology		2.8
Response + more information	12.8	21.2
Insertion sequence + response + more information	8.4	
Response + alternative solution	7.8	9.9
Insertion sequence + response + alternative solution	2.1	
Response + account (+ apology)	13.6	22.9
Insertion sequence + response + account	9.3	
Response + combined actions	8.4	14.8
Insertion sequence + response + combined actions	6.4	

information for the response was added in 21.2 per cent of the calls. Finally, another 14.2 per cent of the calls was made up of responses that are not expanded after the delivery of the news, but their structure is made more complex by the initiation of insertion sequences that delay the actual response. The delay of the response in this format is not made purposefully to take up time before telling the news, but the insertion sequences are initiated to clarify the object of the request, otherwise a response could not be provided.

Table 7.2 will show the frequencies in the use of the different formats by comparing the speakers of the three languages, English, German and Italian. This table shows some differences in the use of the different formats analysed. English, German and Italian speakers can be said to follow different ranking criteria for responding to C's requests for information. If the formats in the cells are considered together, according to the way in which the response has been delivered, the following preferences can be noted. British receivers responded by providing more information next to the response altogether in 24 per cent of the calls, accounting for 14 per cent of responses produced together with an expansion, and another 10 per cent of calls in which the same response

Table 7.2 Actions in the response: cross-cultural comparison (%)

Response format	English (N = 50)	German (N = 42)	Italian (N = 49)
Simple response	4.0	19.0	20.4
Response + apology	2.0	0.0	6.1
Insertion sequence + simple response	16.0	21.4	6.1
Response + more information	14.0	4.8	18.4
Insertion sequence + response + more information	10.0	7.1	8.2
Response + alternative solution (+ apology)	12.0 (2.0)	4.8	6.1
Insertion sequence + response + alternative solution (+ apology)	2.0 (2.0)	2.4	2.0
Response + account (+ apology)	10.0 (6.0)	16.7	12.3 (2.0)
Insertion sequence + response + account	6.0	11.9	10.3
Response + combined actions (apology)	12.0 (6.0)	11.9	4.0
Insertion sequence + response + combined actions (apology)	12.0 (6.0)	0.0	6.1

format is prefaced by an insertion sequence. Another 24 per cent of the calls consists of responses in which the various features for expanding the response are combined. Here, too, 24 per cent consists of 12 per cent calls respectively with the response followed by the extension, and the response with the same format prefaced by the insertion sequence. Consequently, responses extended with an account occur in 16 per cent of the calls, of which 10 per cent is represented by the response produced contiguously to the first-pair part of the adjacency pair, and 6 per cent prefaced by the insertion sequence. Finally, 14 per cent is represented by the extension through the suggestion of an alternative solution, of which only 2 per cent is actually prefaced by an insertion sequence.

In contrast, German Rs more frequently produce responses extended through an account, in 28.6 per cent of the time, of which 16.7 per cent is delivered straightforwardly, and the remaining 11.9 per cent is preceded by the initiation of an insertion sequence. The second most frequent format is represented here, in contrast with the British responses, by the delivery of the response in a simple format, which represents 19 per cent of the cases. In another 20.4 per cent of the calls, the response was delivered in a simple format as well, although it was prefaced by the production of an insertion sequence. The fourth most used response format is the one accounting both for 11.9 per cent of the calls delivered together with the addition of more information, as well as the response extended with a combination of possible actions. The response followed by more information was found to be produced more often after the insertion sequence, i.e. in 7.1 per cent of the calls, than just contiguously to the request, i.e. in 4.8 per cent. In contrast, in the case of the response delivered with an extension through a combination of resources, no insertion sequences were found to be produced when this format was chosen.

Lastly, Italian Rs, like the British ones, extended their responses more frequently with the addition of more information, in 26.6 per cent of the calls, 18.4 per cent of which is produced contiguously to the request, and another 8.2 per cent is prefaced by the initiation of the insertion sequence. They then opted for providing an account of the response delivered in 22.6 per cent of the instances, of which 12.3 per cent was produced straight after the request, and 10.3 per cent was delayed by the initiation of the insertion sequence. Similarly, Italian Rs opted for the delivery of the response in a simple format, i.e. in 20.4 per cent of the calls. It is to be noted that besides the ranking preferences for the use of the different response formats, German and Italian

The Different Response Formats at One Glance 113

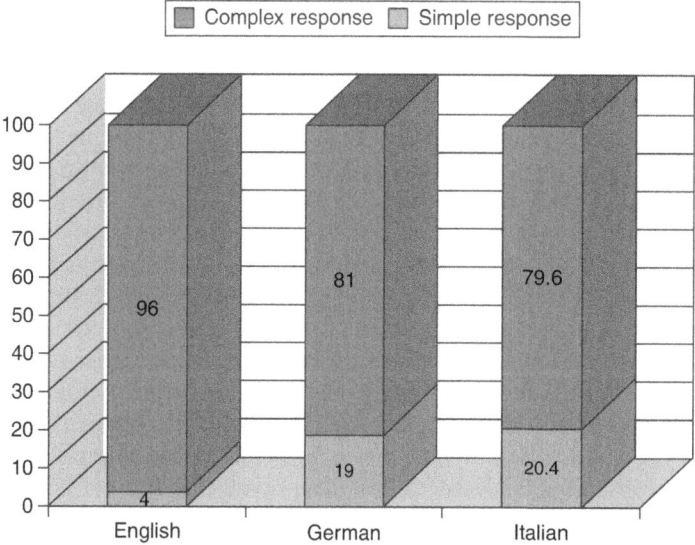

Figure 7.1 Simple vs complex response formats in English, German and Italian

Rs both opted for the use of the simple format of response, accounting for around 20 per cent of the calls in the two subcorpora. As the fourth option for structuring their response, Italian Rs delivered their responses by making use of a combination of the possible actions in 10.1 per cent, 6.1 per cent of which was also delayed by the initiation of an insertion sequence.

What also needs to be stressed here is that the majority of the responses are produced in more complex formats rather than through simple and quick responses in the three languages. Figure 7.1 represents graphically such a tendency to elaborate the response and shows that complex responses are significantly higher than simple ones.

This is especially true for the British data, in which only 4 per cent is made up of simple responses, meaning that simple turn formats are very rare. These results also imply that such features need to be taken into consideration when training personnel working for such services and in call centres. This aspect will be further considered in the following chapter (8), in which implications for training will be discussed.

In some instances the response, whichever way it is organised, is also accompanied by an apology. Apologies were found in 16 conversations overall, representing 11.3 per cent of all calls in the corpus.

These instances represent 8.2 per cent of the Italian calls, and 24 per cent of the British ones. German Rs never accompanied their response with an apology. Apologies were described in the literature as displaying dispreferred features among those strategies that preface the news delivery (Levinson, 1983). The production of an apology in the calls analysed did not systematically preface the news, but was more often produced as a remedial strategy. Its non-systematic and infrequent use is associated only with one type of response, the non-satisfying one. These results contrast slightly with the difference found by Zorzi (1990), Aston (1988a) and Brodine (1991) between Italian and British shop assistants when providing non-satisfying responses. In their analysis they found that British Rs often provided a remedial sequence prefacing the non-granting response, whereas in the Italian data they found such sequences after the news was delivered. Despite such a difference, which is maybe due to the different means of communication employed in the conversations analysed, i.e. the telephone, the study of bookshop service encounters shows similar strategies employed in providing the response to a request for service. Some cases of prefaces to the response were analysed in the previous chapters (cf. example 23, Chapter 6), but the only strategies employed to preface the response were, as already stated, the initiation of insertion sequences, either requests for detail or confirmation checks.

7.3 Turn extensions

There is a final aspect of the analysis to be observed in the corpus, which is the production of turn expansions either as constituents of complex syntactic units, or as free-standing units. These were found to be produced either in the same turn as the core response, or in another. If it was in another turn, the expansion often came after C's production of some receipt of the news through response tokens, acknowledgement tokens, continuers or silences. Table 7.3 summarises the overall occurrences of such expansions of the response, making, as usual, a distinction between the three languages of the corpus.

Responses were incremented in more than half of the calls in the corpus (55.3 per cent) through free constituents, which means the increment was syntactically independent from the core response. Moreover, free-standing increments were mostly produced in turn units separate from the core response, in 35.5 per cent of the calls, compared to 19.8 per cent of increments initiated in the same turn of the response. In the latter case incremented responses made multi-unit turns, composed at

Table 7.3 Free constituents and extension types of the increment of the response (%)

Syntactic format of core response	Simple response	Free constituent		Embedded	Extension	
		Same turn	Other turn		Same turn	Other turn
English (N = 50)	20.0	20.0	40.0	2.0	18.0	–
German (N = 42)	40.5	9.5	33.3	–	16.7	–
Italian (N = 49)	26.5	28.6	32.7	–	10.2	2.0
Total corpus (N = 141)	28.4	19.8	35.5	0.7	14.9	0.7
		55.3			16.3	

least of two units (TCUs): one in which the news was delivered and one with the expansion.

The other type of elaboration of the response, through the production of 'proper extensions', i.e. syntactic developments of the core response, seems to be used much less often, i.e. in 16.3 per cent. As an increment connected to the core response through syntax, this type of extension was found mostly in the same turn of the initial news delivery, and very rarely in a separate turn unit (0.7 per cent), which is only one call. The case in which the extension was produced as embedded in the response, was considered together with the extension type produced in the same turn.

In contrast, simple responses, which are non-incremented ones, represent 28.4 per cent of the calls in the corpus overall. This count is different from that in Table 7.1 above, as here the analysis is based on how the core response was delivered, thus including those calls that receive simple responses even after the initiation of an insertion sequence.

By looking at how the response has been incremented in the three languages, a few aspects need to be noted. Some variability in the way in which free constituents are produced sequentially to expand the response was noted. Italian Rs do not display any difference in incrementing their talk either in the same turn as the core response or in another turn, as free increments are produced in other turns in 32.7 per cent of the calls, and in the same turn in 28.6 per cent, accounting together for 61.3 per cent of the calls. English Rs also increment their responses through a free constituent in 60 per cent of the calls, although the majority of these constituents are produced in a different turn from the core response, i.e. accounting for 40 per cent of free constituents in another turn and 20 per cent in the same response turn. In contrast to British and Italian, German Rs increment their talk through a free constituent less often, in 42.5 per cent of the calls, and similarly to the British they do this in a turn separate from the core response in 33.5 per cent of the calls, compared to 9.5 per cent of the calls in which the free constituent is produced in the same turn. In addition, German Rs are those who tend to produce simple responses more often, in 40.5 per cent of the calls, twice more than their colleagues in Italy and Great Britain, who provide simple responses, even after the initiation of an insertion sequence, in 20 and 26.5 per cent of the calls respectively.

As pointed out in Chapter 4, extensions proper were more often used by British Rs, i.e. in 20 per cent of the calls (accounting for the embedded increment as well), then followed by the Germans, who used this type of increment in 16.7 per cent of the calls. Italians expanded the response in this way, namely through syntactic expansions, half the number of times the British did, i.e. in 10.2 per cent of the calls.

This shows that speakers of the three languages seem to expand the response to a request for service in telephone service encounters, and such an expansion seems to be associated with the overall grammatical configuration of the turn, which tends to consist of simple sentences produced one after the other, sometimes coordinated ones, without being syntactically connected to one another. Moreover, the more frequent occurrence of free-standing increments in turns other than the one of the core response shows this is not just syntactically organised in a simple way, but that it is also cognitively organised at different times. This means that the expansion is not always planned together with the initial delivery of the relevant information about the service. This connection between the interactional organisation of the response and its grammatical configuration will need further analysis.

7.4 Conclusion

The chapter has summarised the results of the analysis with regard to the construction and delivery of responses to telephone service encounters. The description of the various formats of response has shown that Rs can perform several actions while providing the response, and that simple response formats are not the most common way to provide them. As a result, the responses displaying more complex structure represent the majority of the calls, and they can contain up to five different and complementary features next to the actual response: delay of the response through insertion sequence initiation, and its increment by apologising, and/or by giving an account of why the service cannot be provided, and/or by offering an alternative solution, and/or by adding more information about the request.

The summary of the grammatical configuration of turn increments has also contributed to showing which strategies are used by speakers while using the telephone: they mostly expand their responses by adding independent clauses to the initial information, often doing this in turns other than the core response. This result leads to the interpretation that the response and its expansions are planned and produced subsequently and not simultaneously by their speakers. This is also why complex clauses are produced significantly less often.

The next chapter (8) will deal with the practical implications raised by this analysis for business and call centre training.

8
Service Encounters and Call Centre Training Implications

Results found in the deployment of the request–response sequences in telephone service encounters may be useful resources for other fields, first of all education, as they could be used to raise awareness of the practices constituting such encounters.

> The interactional phenomena which are discovered across and within the varieties of settings will enable us to state, with greater certainty, what interactional competencies are requisite to participation in those systems. As such requisites are discovered, we should be able to say what preparation, training, or prior interactional performative skills are vital for new entrants into these systems. And, if members are lacking in particular, identifiable, and describable interactional skills, we should be able to develop methods for teaching, demonstrating, or training those deficient in the requisite skills. (Psathas, 1990, p. 21)

In particular the study of request–response sequences of telephone service encounters may be relevant for training of the personnel of small companies and service providers that respond to phone calls as part of their job, as is the case of Rs of the calls analysed in this book, or who work in call centres. They may be instructed in the proper use of systematic response strategies to the requests for information and pre-requests for service, thus being enabled to offer a better service.

8.1 Inside the call centre organisation

Call centres are contact centres with different roles in that they can either offer customer service, when they are to offer and sell services and quality products, or they offer customer care, when it is their job to have dialogue with customers. Call centres constitute a crucial link between the service providers and their customers in selling products and services for the benefit of the overall impact of the company on the market.

Call centres can have three different possible positions: they can be complementary to a pre-existing commercial structure, they can support it, or they can be an independent selling channel. Call centres are located in almost all business sectors, from telecommunications, to finance, to transport, to health care, to insurance, with different typologies and levels of service being offered:

- Commercial and technical support before and after selling, information about the status of the order, planning of interventions, bills.
- Information about products and services.
- Helpdesks which provide instructions about the use of services and products, their registration, guarantees, technical support, emergency interventions.
- Booking of trips, hotels, flights, etc.
- Buying plans and planning of monthly deliveries.
- Hospital assistance and admissions, health care, ambulatory bookings.
- Selling activity and collection of orders.

Call centres that are complementary to a pre-existing commercial activity or network, have as their primary aim that of implementing the relationship with existing clients in all phases of contact with the customer, and optimising telephone communication by developing 'pro-activity' (Goi, 2005).

Call centres of the third type, which support the selling network, aim at promoting customer loyalty via specific strategies for retaining current customers without getting into conflict with selling activities and assistance. Those centres for selling products are specific and dedicated channels where space is given to the selling as well as to customer service and care.

The call centre represents a special and direct channel between the service provider and its customers. Therefore, the relationship with the customer is of paramount importance, which is why it needs to be analysed to meet the customer's needs. Specific actions need to be taken for successful call centre work.

Two main varieties characterise call centres in the activity they do: 'inbound' and 'outbound' ones. 'Inbound' call centres are those

120 Business and Service Telephone Conversations

dedicated to receiving calls from the customer and they mainly offer assistance. Calls to this type of call centre may aim at requesting commercial information such as prices, product characteristics, selling conditions, promotions, etc., or responding to claims. In contrast, 'outbound' call centres are dedicated to a specific group of actual or potential customers (a reference target established on the base of specific requisites) and aim at presenting a commercial offer, usually in the hope of selling them something. These types of call centre aim at identifying potential customers, presenting the company and the services it offers, and they favour a cross-selling strategy with those who are already customers, plan implementation strategies through a study of actual customers, make enquiries and do market research, and give information assistance and counselling. Besides selling, this type of call centre allows for stimulating interest in possible customers, developing their needs, collecting information and verifying selling opportunities.

The data analysed in this book particularly match the first type of call centre activity, i.e. the 'inbound' one. Although the data do not come from institutionalised call centres directly, they do have to do with customer care and service, and they provide material for reflection about the strategies to be employed by service providers over the telephone.

An important aspect to bear in mind both for small services getting customer calls and structured 'inbound' call centres is the fact that, no matter what the motivation for calling, operators (professionals) should keep in mind that the people calling are comparable to customers entering their shop. In order to sell, shops need people to visit them. The main attraction for getting customers to enter a shop is often having nice shop windows. In the call centre, where eye contact is missing, the shop window is replaced by the R facilitating access to the service, a prompt and quick answer, the homogeneity of the answers given by all professionals in order to guarantee reliability of the service, and the capacity of efficiently analysing a customer's needs and then meeting them.

8.2 Suggested practice for call centres

Manuals for training of call centre professionals stress the importance of communication skills that influence the success of each communicative event (cf. Leidner, 1993; Joyce et al., 1995; Menzler Trott, 1999; Ronchi, 1999; Thieme and Steffen, 1999; Wiencke and Koke, 1999; Finch, 2001; Goi, 2005). Listening is the first of those skills, and active listening is said to be the basis of all these encounters. This implies other skills that impact on the success of the encounter, since by listening professionals can understand the actual needs of the customer, and show such understanding through their future behaviour, i.e. by providing the relevant information or by asking relevant questions in order to better focus on the customer's problem.

Much emphasis is put on the different personalities customers have, and on the professional's capability of trying to adapt to her/his interlocutor's style, be it an analytic, extrovert or direct one. What also have a positive effect are strategies, which show the customer is being properly attended to. Call centre transactions are not purely exchanges of information, but are quasi-conversational exchanges with a customer, and this aspect needs to be taken care of, precisely because this is the distinctive feature of this kind of encounter, which makes using a machine impossible. Notably, not all calls are just requests for information and service, but difficult cases may arise as well, such as when complaints are made. It is in these situations that the professional can put into practice her/his best skills. It is also for these difficult situations that manuals generate lists of rules and guidelines, often with specific techniques for dealing with them. The tendency in this field is in fact that of regulating, and wherever possible, standardising the performance of common conversational routines with the aim of maximising efficiency.

> Left to themselves, operators might design routines that take more time than necessary, or conversely they might aim for speed and neglect other important considerations. These potential problems can be averted by telling operators in detail what to do and say. But to be effective, this strategy must be backed up both by training and regular monitoring. (Cameron, 2000, p. 97)

Standardisation leads to efficiency since by training in the use of different routines in a specific sequential order, operators ensure that the information is elicited in the order in which computers require it to be put in, for instance, so that it can be easily checked for confirmation, and time is not wasted with unimportant repetitions or irrelevant information. What is in danger here, though, in such an attempt to increase efficiency, is a loss of individual language variability if training includes the use of specific phrases together with the adoption of a variety of strategies in order to please the caller/customer. In other words, it may well be possible to perform the same action by using different formulations, typical of the individual style of each R. As Cameron (2000) also

122 *Business and Service Telephone Conversations*

stresses, this attempt at efficiency and standardisation does not necessarily preclude some degree of individual variation in the formulation used. Consequently, providing a set of interactional strategies for the performance of specific actions is suggested, in particular the response to requests for information in service encounters. In what follows, some implications for training will be taken from the data analysed, by providing concrete examples of service providers' behaviour which was either satisfactory or in need of improvement.

8.3 Implications for training: professionalism in the response

Informal interviews with people working in the services involved, which were part of this data collection, were conducted in order to inquire about the level of formal instruction for answering the phone and dealing with customer care and service offers through this means of communication. For the majority of cases Rs reported no formal and specific instruction in the way to respond and deal with customers over the telephone in the specific service where they work. Most of them said they had received general training for working in the service that included some tips on eventual telephone customer care. Only some of the British Rs also reported having had specific training and receiving written instructions with all the resources they have access to when talking with the customer, both on the phone and face to face. This different approach to customer care instruction may be due to the fact that the British services involved, though of a small to medium size like those in the other two countries, are often part of commercial chains spread all over the country that thus standardise their service and approach to the customer.

Such differences may explain the stronger British tendency to expand the response more than 90 per cent of the time, while the German and Italian Rs provide their responses less often (cf. Figure 7.1 and Table 7.2). Moreover, the greater British tendency to response expansion seems to reflect and better satisfy the C's expectations. Cs seem to orient to a professional conversation: there is no need to be abrupt or be in a hurry to reach a result. As pointed out above in Chapter 7, participants in telephone service encounters seem to display haste to get to the business of the call, but then seem to prefer and orient to calm and careful satisfaction of the customer's needs in the body of the conversation. The differences outlined by the analysis of the corpora in the three European languages seem to point out a possible improvement of the service offer and therefore of the training of personnel.

These calls seem to need to follow a different pace in the talk: when they are at the opening phase of the call, it is important to get to the business as quickly as possible; but when they get to the core business of the conversation Rs should orient to an expansion of the response without being in a hurry and by dedicating to such a sequence as much time as is needed. This dedication of time to the core business of the conversation without being rushed appears to be an index of professionalism to which customers orient.

In particular, when getting to the response expansion, the results of the analysis show native speakers (NSs) orienting to a schema for responding to a request for information, that can be used as a referring framework for training. The schema suggests that in order to achieve success on the institutional/business level and maintain a high level of social solidarity with the interlocutor, the response needs to be extended by further talk. As emerging from the participant's behaviour, the response can include:

- pre-expansion through insertion sequences which serve to better tailor the request formulated by C, then followed by a simple and precise answer;
- post-expansion by providing three possible extensions: an account for the response, adding more information to what has been requested, or providing alternatives in response to the request. These different strategies in expansion of the response can also be combined in the response turn, as demonstrated by example 12 (Chapter 4).

The analysis of the responses in the corpus highlighted some tendencies according to which language was spoken. Whereas British Rs seem to tend to combine the different strategies in order to provide a response, German Rs seem to tend more to provide accounts for the response, and Italian Rs to add more information to the actual response. These differences may provide a guideline for training personnel of call centres and insisting more on those aspects that represent a tendency for the speakers of that language in particular, and providing all strategies to be employed in a different order according to the tendency displayed in the results of the analysis of naturally occurring service encounters. As pointed out in one of my previous papers (2007, p. 239) with regard to non-satisfying responses to requests, instruction needs to focus not exclusively on the sequential order in which the responses are formulated, but need to take into consideration the different options for extending the response. There can be two main types of account for not being able to provide a satisfying response:

- The entire category to which the item requested belongs is not among the offers of service provided, and this is made explicit immediately through a statement that prevents all subsequent further questioning by the C.
- The service requested is not available for some reason: the product is out of stock, not yet commercially available, not much requested, not effective, etc.

When alternative solutions to the request are given, in contrast, these may consist of:

- An offer of an alternative item/service by the same provider; this may be a substitute of what the C is looking for.
- An offer to order the item/service requested.
- An offer to check whether a connected shop has the item/service requested.
- The suggestion of a place where the service/item can be found.

In NS talk both the explanation of the reasons why the item/service cannot be provided and the alternative offer accompanying the response are found to be effective but, as shown above, the production of both actions consecutively in the order explanation + alternative solution produces the most effective results in terms of service provider and customer satisfaction. The sequential order followed by NS in providing non-satisfying responses could be used as a template for a more systematic and even standardised way of instructing service providers and staff when a customer's response cannot be satisfied. (Varcasia 2007, p. 240)

The interactional strategies are the same for those calls for which the response fulfils the request in that speakers have been shown to orient to extended responses rather than stick to the simple matter raised by the request. In the following sections some specific aspects of training, common to the three languages observed in the data, will be taken into consideration.

8.4 Specific aspects of training: whose voice has to speak?

Whose voice has to be speak through R's words? Should they speak in the first person singular while providing the response or the first person plural to deliver the information requested? Some manuals for call centre training explicitly remind professionals that they 'are' the brand:

> Specific instructions on how to conduct calls are prefaced by a formula such as 'remember: you are [name of company]'. 'Being' the company means behaving/speaking in accordance with the values it has chosen as central to its distinctive brand image. (Cameron, 2000, p. 100)

Speakers in the data observed switched between the first person singular *I* and the first person plural *we*. In the first instance the use of the first person singular was associated with personal statements of the Rs speaking at that time, showing her/his personal knowledge about the information requested and without judging what was being said, but rather often showing closeness and empathy with C (cf. example 11, Chapter 4). In contrast, the use of the plural is associated with the voice of the company/service provider:

> In institutional contexts, the choice between a self-referring I or we is not 'determined' by the setting; rather, both formulations are available to the institutional incumbent, who can achieve a variety of actions and communicational outcomes by selecting between them. (Drew and Heritage, 1992, p. 63, footnote 27)

Training may therefore point out the different communicational outcomes that the use of one or the other reference produces. Psathas (1990), as well as ten Have (2001), point out a risk in providing training for people doing these kinds of jobs. The risk is that of manipulation of interactional skills and strategies in order to perform the goals of one of the parties.

8.5 Specific aspects of training: apologising

Rs of the three languages observed in this book have employed apologies differently in the response to the request. Apologies in this context do not seem to be a cross-culturally shared convention, but their use seems to be more specific to the British context, since 24 per cent of the calls analysed present it. In these data they do not seem to be associated with specific response formats. As I also report (2007, p. 238), 'apologies were used more often when the response was more elaborate, i.e. when accompanied by an explanation and an alternative offer'. Previous studies on the use of apologies also claim that apologising is culture-specific (Trosborg, 1987;

126 *Business and Service Telephone Conversations*

Olshtain, 1989; Bergman and Kasper, 1993). The results found thus suggest that the apology component of the non-satisfying response by service call Rs is likely to be important for the language instruction of non-native speaker (NNS) staff employed in Great Britain.

In the Italian data, Rs also used apologies to accompany the response, but with lower frequency, i.e. 8.2 per cent. In this case the use of this conversational strategy seems to be linked to repairing a lack of service, almost as a substitute for other types of response extension, as shown in Chapter 4:

> Instruction in Italian would have to consider the different interactional context in which apologies can occur; this seems to be restricted to repairing a response that has been delayed by the production of an insertion sequence and provided with no other kind of extension mitigating the delivery of non-satisfying response. (Varcasia, 2007, p. 239)

Finally, no apologies were used by German Rs. Here the non-satisfaction of the original request seems to be repaired by the use of alternative ways to extend the response, with no need for apologies. It could be suggested that in this language, instruction will need to stress such a tendency for absence of this mitigating strategy, with its apparent inappropriateness in this conversational context for German speakers.

8.6 Specific aspects of training: understanding C's needs

In what follows I will present some examples from the corpus and show some practices. As mentioned at the beginning of this chapter, first of all, the call centre's operator has to show the capacity to understand the interlocutor's needs and problems, which is a consequence of her/his active and attentive listening. Such an attitude is important for the success of the conversation, in that it will guide its flow and provide a basis for the promotion of the service in the future. In the examples that will be presented, some best practice and the resources that Rs have employed in providing the response will first be taken into consideration. Then some instances that need improvement will be focused on in order to point out what common failures can occur, what could be avoided and how they could be improved.

In the first example (25) C's request presents a practical problem: she is asking whether it is possible to buy contact lenses despite having lost the doctor's prescription.

Example 25, Optician, Sardinia (Italy)

```
01  C:   ((telephone rings))
02  R:   arte ottica?                          R: arte ottica?
03  C:   .hh eh buonasera vorrei               C: .hh uh good evening i'd like
         chiedere un'informazione .hhh         to ask for some information
         eh dunque vorrei acquistare           .hhh uh well i'd like to buy
         delle lenti a contatto mensili        monthly contact lenses .hhh but
         .hhh però ho smarrito- oh: il         i lost- oh: the prescription of the
         referto dell'oculista (.) hh (h)      ophthalmologist's prescription (.)
         è po[ssibile                          hh (h) is it po[ssible
04  R:      [e l'occhiale ce l'ha?             R:       [e do you have
                                               the glasses?
05  C:   sì l'occhiale sì l'ho preso           C: yes i do have the glasses
         da voi tra l'altro=                   yes i bought them from you
                                               by the way =
06  R:   =sì va beniss[imo.porta               R: =yes that's ok[ay. you will
         l'occhiale?=                          bring the glasses?=
07  C:                [.hhh                    C:              [.hhh
08  C:   =ah ah?                               C: =uh uh?
09  R:   noi leggiamo                          R: we read from the glasses
         dall'occhiale::
10  C:   la gradazione                         C: the dioptre
11  R:   sì (per prendere) la gra-             R: yes (to take) the dioptre
         dazione
12  C:   mh h okey grazie e:?                  C: mh h okay thank you u:?
13  R:   prego                                 R: you're welcome
14  C:   arrive[derci                          C: arrive[derci
15  R:         [buonasera                      R:       [goodbye
                                               (GDI 19 *NU24F1 *SS OTTICA)
```

R shows her understanding C's problem by promptly asking for some clarification before C has actually got to a completion of her turn (line 04). Once she gets the response, she promptly proceeds to providing the solution in the response in line 06 (i.e. *porta l'occhiale? You will bring the glasses*), followed by an expansion with a more specific explanation about the procedure to follow in that case (i.e. in lines 9–11, *noi leggiamo dall'occhiale:: [...] per prendere) la gradazione, we read from the glasses [...] to take the dioptre*). In this instance we can also see how the expansion of the response is also collaboratively completed by C (lines 09–12).

Efficiency of the response is given here not just by the prompt clarification request that helps R focus on the request and its subsequent response, which is then delivered together with an explanation of the procedure, as we have just seen. Efficiency is also given by R using positive assessment markers in the response (i.e. in line 06 *sì va benissimo, yes that's okay*), which upgrade its value and convey the positive attitude operators are suggested to have in order to show the nice 'shop window' of their company.

In the next example C calls the health care offices to ask for some information about what she has to do in order to get health care assistance while being abroad for a short period of time.

Example 26: Health care office, Sardinia (Italy)

01 C: ((telephone rings))
02 R: °pronto°? — *R: °hello°?*
03 C: eh:: buongiorno .hh eh:: mi scusi io (.) mi serve un'informazione. — *C: eh:: good morning .hh eh:: excuse me i (.) i need some information.*
04 R: °mi dica°. — *R: °tell me°.*
05 C: .hh eh:: allora sono una::hm una studentessa .hh eh:: devo andare in viaggio in inghilterra. — *C: .hh eh:: well i am a::hm a student .hh eh:: i have to go to england for a trip.*
06 R: °sì (.) mi dica°. — *R: °yes (.) tell me°.*
07 C: vorrei sapere (.)eh::hm per un mese come p- come devo fare per l'assistenza sanitaria. — *07 C: i'd like to know (.)eh::hm for a month what c- what do i have to do to access health care.*
08 R: un mese(.)solamente? — *R: one month (.) only?*
09 C: sì per un mese. — *C: yes for a month.*
10 R: niente:: viene qua col libretto sanitario- oh[:: — *R: nothi::ng you come here with the health card- oh[::*
11 C: [sì — *C: [yes*
12 R: e noi rilasciamo il modello e centoundici. — *R: and we fill in form one hundred and eleven.*
13 C: ecco — *C: that's it*
14 R: °e nient'altro° — *R: °and nothing else°*
15 C: e de- devo compilare questo modello oppure? — *C: and do- do i have to fill in this form or?*

16	R: no no no lo compiliamo noi? [il modello	R: no no no we will fill it in? [the form
17	C: [ah ecco e così io sono coperta per quel mese:	C: [ah ok and so i am covered for that mo:nth
18	R: (esatto)	R: (exactly)
19	C: va bene va bene la ringrazio.	C: that's okay that's okay thank you.
20	R: (°buongiorno°)	R: (°goodbye°)
21	C: arrivederci	C: arrivederci

(GDI 11 *NU40F1 *NU ASL)

As in the previous example, R responds to the request turn with an insertion sequence in which he asks for confirmation. His response comes in the following turn by minimising the issue to C with the turn beginner *niente*, *nothing* and then explaining the steps to take. The response is then expanded in line 14 in uptake to C's reaction to the response just delivered by again minimising the supposed complexity of the request posed by saying *e nient'altro*, *and nothing else* (line 14), thus implying that the procedure to follow is very simple and quick. This is followed by C's clarification request in line 15 about the form she should get from the office and what she has to do in order to get it correctly filled in.

Both examples show how speakers orient to the extended format of response comprising the initiation of a short insertion sequence prior to the response to C's supposed complex and problematic request for information, which in turn, in order to take care of the customer at the other end of the line, provides the relevant information as requested together with an extension. In addition, both Rs deliver the message in their responses that the issues posed by C in the requests are not as problematic as they were apparently conceived initially, and they do this by delivering the response with positive assessments in the first case (example 25), and minimisation markers in the second (example 26).

In the following example, a call to a bookshop, C asks about the availability of law books, and R's response is quite different in terms of strategies of response with respect to the previous example. R immediately replies with a no, *no signora*, *no madam* (line 04). In the same turn, just after having delivered a negative response, s/he asks for more information about the books in terms of use and purpose, then offering a possible candidate, i.e. *per dei concorsi?*, *for exams?* In doing this,

R shows a lack of understanding of the matter of the request and the late production of the clarification request conveys low willingness to understand the customer's needs.

Example 27: Bookshop, Sardinia (Italy)

01	C:	((telephone rings))	
02	R:	°libri scolastica°	*R: ° scolastica books °*
03	C:	e:: buonasera senta avrei necessità di un'informazione, avete dei testi giuridici?	*C: e:: good evening (listen) i need some information, do you have law books?*
04	R:	(no signora) ma di che genere? pe:r dei concorsi?	***R: (no madam) but what kind? fo:r exams?***
05	C:	>no no no< i codici e:: i manuali di diritto e di procedura civile e penale	*C: >no no no< the statute books a::nd the law manuals and those of civil and criminal proceedings*
06	R:	no non li abbiamo	***R: no we don't have them***
07	C:	la ringrazio. buonasera	*C: thank you. good evening*

(CVI 8 *NU42F1 *SS libreria2)

Once C clarifies what she means by 'law books' in the following turn (line 05), R replies simply by confirming the negative response. Here the response is accompanied neither by an apology nor by any directions about where such books can be found, thus being general and synthetic.

In this case there are a few aspects for improvement of the response given by R. First of all, when the call centre operator does not know the response to C's request because it sounds unusual with respect to the normal range of requests received, s/he has to try to avoid anticipating a negative response. In other words, never start your response with a *no*, especially if you are not sure about what you are talking about. Second, rather formulate a clarification request before giving the actual response, so that you can be sure of having properly understood C's needs, and in this way also show the recipient you understand her/his request. Lastly, when delivering the response, if you are not able to provide more information through an account that explains the reasons why the service is not available or provide alternative solutions, you can accompany your response with an apology that with a minimal expansion of the turn of talk maintains high social solidarity with C.

The call centre operator has always to keep in mind that one of the most important aspects of her/his job is that of trying to understand C's problems and issues and show such understanding by trying to provide the right solution.

The next example is a call to a wine shop in the UK in which C asks about the availability of a specific type of Italian wine.

Example 28: Wine shop, Portsmouth (UK)

```
01  C:  ((telephone rings))
02  R:  good afternoon wine spot (    ) cristoph speaking
03  C:  .hh hello i was wondering could you tell me do you stock
        any sardinian wines
(.)
04  R:  sorry?
05  C:  do you stock any sardinian wines
(0.6)
06  R:  >what is that i don't know<
(.)
07  C:  e:m wine from sardinia
(1.2)
08  R:  wine from
(.)
09  C:  sardinia (.) in italy
10  R:  no
11  C:  no? [do you know anywhere=
12  R:       [no i don't have it
13  C:  =i could get it from?
14  R:  oh i'm new in the business? i can't help you i'm sorry
15  C:  ah okay that's no problem thanks very much
16  R:  sorry bye
17  C:  bhm bye bye >no problem<
```

(SNGB 07 *IR24F1 *POR WINES)

In this example, R begins by claiming his ignorance patently (i.e. line 6, *what is that I don't know*) in response to C's request. The same message is conveyed in the following turns through the delayed initiation of the clarification request through a partial repeat in line 8. And when the clarification comes in line 09, the response is delivered abruptly with a *no*. But the conversation has not yet come to an end. R's following turns do not improve the situation when R insists on responding that he does not have the requested product after C echoes his response as news receipt. Then she starts her request for a suggestion of another place that might have it (lines 11–12), to which R responds in a non-collaborative way by admitting that he is new to the business.

This example also offers a few instances for improvement of the response provided. First of all, never say overtly that you do not know what the customer is talking about. If this is the case, R and call centre operators should always try to ask for clarification about the product requested without admitting their ignorance. In saying they do not know, operators discredit their company. Further, and as mentioned previously, avoid formulating the response by beginning with *no*. Thirdly, avoid saying you cannot help, as R does here in line 14. Such a move also discredits the service provider and excludes it from the market.

The last aspect to bear in mind is that of keeping the conversation to an appropriate length. In the following example C rings to ask for information about package holidays to Spain (line 05). R's first response in line 06 certainly shows agreement and provides a positive response to C's question, though it lacks a focus on, and restricts the general request for information to, a specific aspect to be deployed. C responds to R's reaction to her question by getting to the point in lines 12–13 step by step.

Example 29, Travel agency, Sardinia (Italy)

01 C: ((telephone rings))
02 R: pronto?   R: hello?
03 C: e: pronto buongiorno   C: e: hello good morning could
   potrei parlare co:l   i speak to:
   signor caravag[gio?   mister caravag[gio?
04 R: [sono io signora [buon-   R: [it's me
   giorno   madam [good morning
05 C: [.hh   C: [.hh good
   buongiorno ascolti volevo   morning (listen) i wanted to
   sapere se organizzate   know if you also organise
   anche: viaggi: di studio in   study trips to spain
   spagna
06 R: sì certamente   R: yes certainly
07 C: sì:?   C: ye:s?
08 R: sì sì sì sì sì   R: yes yes yes yes yes

09 C: ecco a me interessa mm
        più o meno il periodo di
        agosto
10 R: certo
11 C: e:: volevo sapere se
        sono ancora in tempo
        [oppure:: =
12 R: [sì sì
13 C: =è tardi
14 R: be? ci mancherebbe [noi
        andiamo avanti=
15 C: [mh
16 R: =fino ad agosto poi con
        le prenotazioni. (.)
17 C: ho ca[pito
18 R: [praticamente:: è a ciclo
        continuo.((ride)) chiaramente,
        il discorso è questo,
19 C: sì
20 R: prima si prenota? e
        meglio è? perché. perché
        così riusciamo a trovare? (.)
        anche più disponibilità per
        quanto riguarda l'alloggio
[...]

30 R: con la spagna? è un po'
        più difficile. dove le inte-
        resserebbe in spagna?

31 C: ma e:: non ho un e::
        diciamo una meta pre-
        cisa. io? (.) studio lingue
        [e::
32 R: [>ho capito. ha visto la
        locandi-< ha saputo:: da:
        tramite:: ha saputo di
        noi tramite qualcuno?
        che [è partito con noi

C: well i am interested mm
more or less in august

R: sure
C: a::nd i wanted to know
if i am still in time [or:: =

R: [yes yes
C: =is it late
R: well? Not at all [we go
on =
C: [mh
R: =until in august then
with the bookings. (.)
C: i se[e
R: [in fa::ct it's a never
ending cycle. ((laughter))
obviously, this is the thing,
C: yes
R: the sooner one books?
And the better it is?
because. because this way
we manage to find? (.) also
more availability regarding
the accommodation
[...]

R: spain? that's a little
more difficult. Whereabouts
in spain would you like to
go?

C: well e:: i do not have
a specific destination in
mind. I? (.) study languages
[e::
R: [>i see. Have
you seen the poster-< did
you get to know about us
through somebody? One of
our clients

33 C: [sì tramite miei colleghi che::	C: yes, my colleagues have told me tha::t
34 R: ah	R: ah
35 C: sono partiti con voi sì	C: they were your clients yes
36 R: ((cough)) chiedo scusa	R: ((cough)) sorry
37 C: e: niente	C: e: that's ok
38 R: ((continues coughing))	R: ((continues coughing))
39 C: e quindi: appunto. la mia: non ho una destinazione precisa por- mi interessa appunto trascorrere un periodo lì e::m magari frequentare una scuola	C: and so: then. My i don't have a specific destination por- i'm just interested in spending some time there u::m maybe attending a school
40 R: be certo certo	R: well sure sure
41 C: e:: quindi avere insomma? non so? andare in famiglia?	C: a::nd then having well? I don't know? staying with a family?
42 R: certamente? guarda, tanto ci possiamo dare del tu visto che tu se[i:: >sicuramente< più giovane di me=	R: certainly? look, call me [first name] since you ar[e:: >certainly< younger than me=
43 C: [(((ride)) e:: sì sì	C: [(((laughter)) u:: yes yes
44 R: io ne ho trentatre tu quanti ne hai scusami?	R: i am thirty-three you how old are you sorry?
45 C: e: ventidue	C: u: twenty-two
46 R: e:: allora dai [dammi del tu ( )	R: u:: then come on [call me [first name] ( )
47 C: [e: sì ((imbarazzato))	C: [u: yes ((embarrassed))
48 R: senti:: come ti chiami mi hai detto?	R: listen:: what did you say is your name?
49 C: e:: sandra	C: e:: sandra
50 R: sandra. allora guarda? per la spagna abbiamo tutto, ma proprio tutto tutte le destinazioni possibili e immaginabili (.) allora barcellona e::m madrid e:	R: sandra. So look? for spain we have everything really everything all the destinations you can imagine (.) well barcelona e::m madrid e:
51 C: mh mh	C: mh mh

52	R:	sivi- e valenzia siviglia e:: d- salamanca
53	C:	sì
		[...]
84	R:	ascolta? [e: se tu hai=
85	C:	[sì
86	R:	=la possibilità di fare un salto? [da noi
87	C:	[certo? magari mi avvicino per avere ulteriori informa[zioni
88	R:	[ecco. anche stamattina se vuoi? verso: fine mattina:[ta quando vuoi tu=
89	C:	[mh eh?
90	R:	=quando preferisci
91	C:	stamattina non so se posso. magari domani mattina mi avvicino=
92	R:	ecco? io ci sono dopo le <undici> e mezzo >la mattina mi trovi sicuramen[te<? e la sera=
93	C:	[ho capito
94	R:	=dopo le cinque e un quarto cinque e mezzo anche fino a tardi.
95	C:	mh mh?
96	R:	la mattina? ma anche fino all'una e mezza almeno ci sono sicuramente
97	C:	va bene
98	R:	d'accordo?
99	C:	okey ti ringra[zio
100	R:	[bene >sai dove siamo allora< sì?
101	C:	sì sì [mso::
102	R:	[su al >centro commerciale luna e sole<

R: sevi- e valencia sseville
e:: d- salamanca
C: yes

R: listen? [and if you have =
C: [yes
R: =the possibility to drop in? [here at the office
C: [right? I could pop in to get further informa[tion
R: [right. Even this morning if that suits you? later this morn[ing when it suits you=
C: [mh eh?
R: =whenever you want
C: i do not know if i can make it tomorrow morning i can come=
R: right? I will be here after <eleven> thirty >you will find me in the morning? and in the evening
C: i see
R: =after a quarter past five and half past five also later.
C: mh mh?
R: in the morning? But also until one thirty at least i will certainly be here
C: good
R: okay?
C: okay than[ks
R: [right >do you know where we are< yes?
C: yes yes [mi kno::w
R: [in the >mall luna e sole<

103	C:	sì lo so.mh. gra[zie sì	C:	yes i know.mh. tha[nks sì
104	R:	[d'accordo, cia:[o	R:	[right, cia:[o
105	C:	[arrivederci	C:	[bye

(SNI 15 *SS CVS)

The call is made to the same place as in example 13 (Chapter 4). It goes on for three minutes with R providing various items of a list of general information about the possibilities the agency offers for students wishing to go abroad for some time, talking about other students who have travelled with their offers, and trying to make contact with the student, by asking questions which are not relevant to the request, but rather pertaining to his personal interest (i.e. line 32, *ha visto la locandiha saputo:: da: tramite:: ha saputo di noi tramite qualcuno? che è partito con noi*). Such behaviour should be avoided by a call centre operator for at least two reasons: no matter who C is, it is not very professional to delay the call by not focusing on the real issue raised by C. Similarly to what happened in example 13 above, there is no reason for keeping C on the phone, for just chatting. It is R's job to answer C's issues, but not to start chatting.

The last aspect raised is also reinforced by another action performed by R in this call at the sequence initiated in lines 42–46, in which R initiates a side sequence asking C to shift from the formal '*Lei*' to the familiar '*tu*'. This is done by justifying that C is certainly younger than R, who describes himself as young. Besides being off the topic of the conversation initiated by C, such a request for a shift displays another aspect of poor professionalism, since it is not up to R and the call centre operator to make such kinds of requests. Such examples are inefficient and do not comply with the standardisation mentioned above. Although R shows a positive approach, which is desirable for all call centre operators, his lack of training makes him perform the wrong actions throughout the call, by losing efficiency in keeping the conversation unnecessarily long. Additionally, asking for personal information for no purpose and switching to an informal register are risky.

8.7 Service encounters and their implications for call centre training

Call centres, as was also observed by Cameron (2000), are of particular interest because of the attempt by training manuals to regulate many aspects of talk, often trying to get it to be automatic. Reality shows that behind the apparent ease of the telephone transaction, many aspects actually need attentive scrutiny and reflection before becoming talk. The examples observed here, coming from a closed setting, that of incoming calls to small service providers all over Great Britain, Germany and Italy, have shown only some of the complex aspects that should be taken into consideration for the success and improvement of those interactions.

Call centres attempt to regulate talk in order to maximise efficiency, and predictability and even 'calculability' come into play. Such attempts at rationalising talk occurring in call centre transactions conflict with one of the foundations of the work in CA, i.e. that talk is a 'locally managed' phenomenon. This means, as introduced in Chapter 1, that the rules and procedures underlying communication are applied by participants as they go along the interaction in order to shape it.

> In the call centre regime, by contrast, attempts are made to manage talk globally: to stipulate in advance how long an interaction will last, what moves it will consist of and what the outcome will be. As Drew and Heritage (1992) observe, some degree of global shaping is a feature of many kinds of institutional talk, but in call centres it is taken to such an extreme, it is hardly surprising that problems arise from it. (Cameron, 2000, p. 123)

With exactly this aspect in mind we have made the suggestions in this chapter on the basis of the data observation and analysis, aiming at contributing to the description of the different sequential moves that service providers can employ in their telephone encounters for the accomplishment of the request/response sequence. This can be part of the set of communication strategies call centre operators can reflect on and be trained in.

9
Conclusions and Implications

This book has tried to contribute to the study of sequential organisation of the request–response adjacency pair in service encounters and to an understanding of the cross-cultural similarities and differences found among the three languages under consideration. All such findings may have implications for improvement of customer service in small businesses and the training of assistants, clerks, call centre operators and learners of the three languages, as has been discussed in the previous chapter (8).

This final chapter will look at some of the issues that the analysis of telephone service encounters raised and that can be taken into account in further research. Firstly, service encounters will be observed in their institutional framework as constituting a specific conversational genre, with its rules and routines. Secondly, the chapter will also raise some implications for the methodology used, such as the use of both qualitative and quantitative methods of analysis in CA studies and the conceptualisation of analytical concepts in this field of research, some of which have been found to be non-exhaustive and ambiguous. The implications for the use of data from different types of services and cross-cultural comparison will also be considered.

9.1 Service encounters as a genre

Participants in the service encounters analysed have been shown to incrementally accomplish (Schegloff, 1982) the discourse they are engaged in by following shared schemas and orienting to them moment by moment in the interaction. Most of these features of talk-in-interaction in telephone service encounters seem also to be cross-linguistically shared among speakers of British English, German and Italian, with the preferences for one format or another outlined in Chapter 7.

Conclusions and Implications 139

The analysis started from a hypothesis that needed to be confirmed by the participants' actual orientation: that participants would orient to brevity and time pressure as they do in the opening sequences of service phone calls. Such participant orientation in the opening sequence is also pointed out in several studies and accounts for several languages: British and American English, German, Finnish, Greek, Japanese, Italian (Zimmerman, 1992; Bergmann, 1993; Halmari, 1993; Pavlidou, 1994; Wakin and Zimmerman, 1999; Bercelli and Pallotti, 2002; Luke and Pavlidou, 2002; Thüne, 2003). These contributions to the study of call openings all stressed the production of the core opening moves compressed into a few turns, in contrast to the openings of calls made between acquaintances. But the sequences beyond the opening, i.e. those that deal with the reason for the call, seem to outline a different orientation of their participants. The analysis of such sequences in the previous chapters has shown that here speakers seem to orient more to the business details, taking the call as an opportunity to make money. C then becomes a possible customer who needs to be pleased by using all possible resources available to R to provide the service requested.

This concern with selling means that participants do not orient to brevity in the responses analysed, but they rather prefer more complex response types and more talk in general, so that even if the response displays the availability of the service requested, this cannot be confined to a simple '*yes we do*' response type, because it would sound abrupt and impolite. So, in the corpus analysed Rs tend to extend their responses as has been described in the above chapters, and Cs expect this in the way they provide receipts to the responses.

Speakers thus follow predefined response patterns that can be traced back to a specific communication genre. The response pattern to the reason for a call could be summarised as:

1. Responding by saying the relevant information first: 'yes you do, no you don't' offer the service requested;
2. Then extending your response, by giving an account of why you deal, do not deal with that service or providing an alternative solution;
3. If the request made is not clear, asking the relevant questions to narrow it down and then responding following the schema in steps 1 and 2;
4. Trying to avoid abruptness and trying to leave your interlocutor with more information than has actually been requested.

The continuous handling and adjustment of the talk to the context in which it takes place therefore lead to a slight modification of the notion of public service encounters (PSEs) outlined by Pixi project researchers for bookshop encounters:

> Our notion of a PSE is that of an encounter which initially matches a PSE schema, where the discourse confirms the basic expectations it sets up for a 'business transaction' as the initial talk-type. This schema, stating typical situational features and associating these with certain typical discourse ones, may be referred to by participants in the discourse process. (Aston, 1988b, p. 40)

Moment by moment, the schema is taken into consideration by the interactants; its reference is not just a possibility which participants can access during their conversation, but something speakers refer to continuously, and according to it they adapt its shape (cf., for instance, example 12, Chapter 4).

9.2 Methodological implications

The analysis of service encounters in three different languages by using CA as a methodological and theoretical approach raises various implications, some of which will be dealt with in the following sections. In particular, the use of both qualitative and quantitative analysis in the observation of the data will be discussed, and by using an approach, suggested by CA, that is mostly based on careful and detailed qualitative analysis. The chapter will also deal with a problematic issue which has already been raised in Chapter 1, which is the lack of systematic operationalisability of some analytical concepts theorised in the CA paradigm. The issues raised by comparing various types of service encounters, instead of just one type, such as just bookshops, or travel agencies, or museums, for instance, will also be discussed. Finally, the cross-cultural comparison will be dealt with.

9.2.1 Combining qualitative and quantitative methodologies

The use of quantification in CA, together with qualitative and detailed analysis of 'what's going on here' (Goffman, 1981), may ground the assumptions raised from just qualitative studies on empirical evaluations. Its use, as Schegloff (1993) points out, cannot replace the detailed observation of interactional sequences, but certainly represents a further aid to researchers in this field. Access to larger corpora quantification represents a decisive step in evaluating the use of certain conversational strategies and providing evidence of their role in the exchanges (Drew, 2005).

By carrying out quantitative analysis and attributing categories to the phenomena, observation of the data may certainly lose in richness of detail, as the researcher and the reader abstract from the data. The repeated observation of a phenomenon leads to inferences about its use in one context or language more than another. Such repetition in the occurrences of the phenomenon makes them belong to the same category, so that they are then abstracted from their individuality (cf. Colamussi and Pallotti, 2003) and the individual peculiarity of the action becomes blurred.

In contrast, quantifying phenomena that are central to the interactional organisation and development of the conversation provide an account for its use over a population. In addition, such results represent important information for practical application of studies in this field. So what matters is not just the discovery of a phenomenon and an action, but also their more or less shared use by a group of people, who are representative of a social group. Social practices, and ways of performing them, need to be analysed and pointed out even when they are produced only once, because they represent the occurrence of a phenomenon that belongs to the variety of actions that can be performed. This is why quantification should represent the very last stage of analysis, after a previous and careful observation of the data, which reveals what is going on among the participants in the interaction. Quantification in CA, as in other disciplines, can then be a precious fine-tuning device for grounding and extending generalisations on the distribution of specific phenomena in the conversational exchange, alongside the usual analytical procedure used in this field.

9.2.2 Definition of analytical concepts

Chapter 1 pointed out the advantages and merits of conversation analytical concepts in understanding talk-in-interaction, and it also raised some problems of systematic applicability and practical use of the concepts initially theorised by CA pioneers. In particular, what happened in the spread of the discipline is that the concepts initially theorised were found to be unclear in their formulation, which leads to multiple interpretations, sometimes misconstruing the actual meaning and content of the concept. A reason for such confusion is probably to be found in a certain lack of methodological manuals for instructions on how to do a good analysis of talk-in-interaction. Only in the last two decades has work in this direction been done, and some useful publications have made accessible all the methodology employed by conversation analysts (Psathas, 1995; Hutchby and Wooffitt, 1998; ten Have, 1999; Drew and Heritage, 2006; Schegloff, 2007; Sidnell and Stivers, 2012). But here, too, some of these concepts could not be explained extensively if there was a lack of explicitness or wide applicability in the initial formulation of the concepts. Another reason may be found in the focus of CA studies on specific and detailed phenomena in specific contexts. This method has both positive and negative impacts. On the one hand, careful analysis of constrained events reveals their specific orderliness in the organisation of talk. On the other hand by providing a description of the strategies used in constrained events, they cannot be easily extended to other contexts, unless more work is done. This process is very slow.

During the analysis of the data in this book we came across some of these unclear concepts such as TCUs and turn extensions. The reasons for such confused understanding of the concepts and their applicability may vary. TCUs (Sacks et al., 1974), for instance, were found to be formulated with some unclear features. They did not make clear the conceptualisation of units for large stretches of talk (Selting, 2000).

This section certainly does not aim to accuse previous and eminent researchers of bad conceptualisation. CA is a fairly young discipline in the study of social interaction. Therefore more studies on various conversational events are needed that may then contribute to adjusting the definitions of the analytical concepts used, by confirming or disproving the conventional formulations. What is necessary is a clear definition of each concept that makes explicit all features belonging to the concept and also lists what phenomena are not included.

9.2.3 Comparing different types of services

The data analysed are representative of various service providers in the three languages English, German and Italian, as described in Chapter 2. Previous work has focused on distinct types of services in the study of talk-in-interaction, such as bookshop encounters, as in the Pixi project (Aston, 1988a; Zorzi, 1990; Brodine, 1991; Gavioli, 1995), or the reprographic store drop-off counter (in Vinkhuyzen and Szymanski, 2005), or supermarket interactions (De Stefani, 2006, 2008), or the calls to an airline company (Lee, 2011a, b), or emergency conversations (Whalen and Zimmerman, 1987; Zimmerman, 1992; Zorzi, 2002; Monzoni and Zorzi, 2003; Zorzi and Monzoni, 2004), and so on. The focus on just one type of service has clear advantages. The variables that come into play are much more controlled with respect to instances when more than one business is considered. The types of actions that may be performed will be constrained and found to be typical of one type of encounter, as well as that each institution may have specific strategies for reaching their aim with success and dealing with the customer. In contrast, the focus on various service providers may show that they share both the same actions and the production of common strategies to perform them. This was the hypothesis in the present study, which led to the collection of talk-in-interaction taking place in different services. The study has then confirmed our hypothesis, confirming that request–response sequences do not vary much from service to service. Participants were found to share common strategies about how to organise a response to a request for service.

In the observation of more than one service type the researcher takes a different perspective. S/he needs to focus on just one sequence type or phenomenon, and see whether this is shared by all institutions taken into consideration. This is what happened in the request–response sequence taken into consideration here. The focus of the study is then constrained to just one phenomenon and looks at its deployment. Such an approach may also be taken when looking at just one type of institution. Looking at the deployment of an action in various services leads to a wider observation of the world, but still shows that different strategies are used to deal with and solve problems with the customer from one service to the other.

The study of more than one service is more complicated, as the corpus needs to be built carefully. For instance, not all institutions are easy to compare, so that in this study the choice fell on the interactions in small and medium-sized businesses.

Another practical implication for setting up the present study concerned the collection of the data. Systematic and ethical data collection raises the issue of spontaneity in the production of talk. Participants were then provided with a list of services that had given their consent for recording and they could call the services they mostly needed to get in touch with.

9.2.4 Cross-cultural comparison

Cross-cultural comparison is a practice that attracts many researchers for its practical implications on intercultural communication. The aim is well known, i.e. that of looking for similarities or differences in talk-in-interaction. When differences are found, they are usually in the way an action is performed, although the main sequential rules are shared. Here the researcher's task is that of finding out the consequences of such differences (Schegloff, 2002, p. 250). According to Schegloff, though,

> in exploring such differences as are found, the focus has been not so much on the consequences of the differences for the interaction itself as on the differences as indicative or symptomatic of divergent themes and features of the larger cultural context – which is quite a separate undertaking (ten Have, 2002). What has happened in this literature is that openings have been disengaged from the conversations which they were opening – and which they were designed by their parties to open, and have been juxtaposed instead with other openings, drawn from different cultural settings. (Schegloff, 2002, p. 250)

Schegloff thus says that often the cross-cultural differences pointed out by researchers, in particular by comparing phone openings in various languages, have been abstracted from the relevancies of the *participants in the conversations in question*, but were the result of thoughts of academicians, looking for cultural differences in their profession. When comparing two or more cultures in the deployment of specific conversational events it is therefore important to let the participants talk from the data rather than the researcher's thoughts and intuitions.

The comparison of the three languages taken into consideration for this study, English, German and Italian, showed that, on the whole, the resources used in the development of the request–response sequence are shared. This type of analysis, though, does not trace absolute cross-cultural barriers and bridges among the three languages. The analysis of cultural phenomena is complex, and it is hard to say that the data analysed can represent the culture of the three languages. The results found are certainly a trend in the samples analysed, which the reader may well consider arise from the general cultural context. From what can be seen from the data, there are shared strategies and ways of dealing with the response in a telephone service encounter in the three countries, with slightly different preferences for one or the other response format described.

9.3 Conclusion

By raising such implications both for the methodological and practical level this study hopes to contribute to the present research on talk-in-interaction and its usability in other fields. As in most research, this book has contributed to describing what strategies are employed in providing a response to service requests that are consistent with both the speakers' expectations about the kind of conversation they want to have. At the same time it has left some issues unresolved and raised new aspects that could be studied further.

Appendix

Here follow the complete texts of the transcripts of the calls analysed. They are organised according to the language in which they were produced and are presented in the order they are presented in the chapters.

English calls

SNGB 04 *IR24F1 *LO BEAUTY CENTRE
01 C: ((telephone rings))
02 R: (an absolute look) how may i help?
03 C: .hh hello i was wondering could you tell me how much it costs fo:r a manicure please
    (0.3)
04 R: a manicure?
05 C: mh[m
06 R: [fifteen pounds
07 C: okay >and do< i have to make an appointment
08 R: (we wont make an appointment) but if you come in when i'm busy (then i will fit you in)
09 C: .h ah okay that's great thanks very much?
10 R: you're welcome
11 C: bye bye
12 R: bye

SNGB 13 *IR35F1 *POR ESTATE AGENT'S
01 C: ((telephone rings))
02 R: thank you (   ) douglas susanne speaking how can i help?
03 C: .h good morning susanne. em i was em wondering e:m if you: also: (.) do things do with apartments (.) leasing apartments and thash
04 R: we do have a letting department
05 C: you do have a letting department.
06 R: yeah. >i can give you the number< it's actually a separate office [e::m
07 C: [°okay°
08 R: it's imp- it's a north end it's o two three nine two.
09 C: o two three nine two.
10 R: six five three
11 C: six five three
12 R: one five seven
13 C: oke[y thanks=
14 R: [okay?
15 C: very much
16 R: not a problem. [b-bye
17 C: [bye-bye

146

Appendix 147

GDGB 1 *POR50M1 *POR BOOK SHOP
01 C: ((telephone rings))
02 R: °good afternoon all^books ?°
03 C: .hh oh good afternoon i'm interested in the new harry potter book .hh i wasn- are you taking orders for (it)(so) when does it c:ome out.
04 R: .hh there's still no publication date i'm afraid (0.5) a:nd so i don't think °we can actually place an order cause then we wont be able to give an isbn°.
05 C: i see . so i have to leave it a while.
06 R: yeah. [( )
07 C: [ok.
08 R: i'm after it myself but? [you know=
09 C: [yeah
10 R: =we- i think (0.2) leave it an other mmonth (0.2) and there we would-should be able-we-as soon as e: they give us a publication date they'll give an isbn number as well?
11 C: okay
12 R: and then we can place °your order°.
13 C: thank you very much
14 R: okay?
15 C: yeah
16 R: [°thank you°
17 C: [bye
18 R: bye

GDGB 10 *POR29M1 *POR BOOK SHOP
01 C: ((telephone rings))
02 R: good afternoon all^book('s).can i help you?
03 C: yeah. i'm just making an enquiry do you have hm (.) i don't know if it's a biography or an autobiography for victoria beckham? .hh do you have it in st[ock?
04 R: [no no we don't stock ehm general biographies unfortunately.
05 C: right.
06 R: (we could order) to you one. e:m do you want it today?
07 C: .hh no it's i- i- i- it's not necessary, no. [i were=
08 R: [no
09 C: =just (.) doing an enquiry.
10 R: okay. no. we don't actually have one in stock, but i can certainly research one for you (   )
11 C: e:m i've got-i've got a hurry at the moment. but e:m i'll phone you back later. [°okay?°
12 R: [okay a: well we're closing in fifteen minutes.
13 C: okay.
14 R: okay? yeah? e we do have a web site if you have got access to the internet?
15 C: okay, what's your website?
16 R: it's www dot
17 C: yeah?
18 R: allbooks? [dot=
19 C: [yeah?

148 *Appendix*

20 R: =co dot uk.
21 C: okay.
22 R: all right?
23 C: thank you. bye?
24 R: bye-bye

CVGB 17 *LON33F1 *LCS Museum 1
01 C: ((telephone rings))
02 R: good afternoon paper museum?
03 C: .hhh hello i was wondering if you could tell me e:m what are the opening hours fo:r the weekend for the mus-museum
04 R: ehm (.) the paper museum?
05 C: .h yeah
06 R: e: we're open everyday ten till five apart that we're not open on a sunday
07 C: .hh not open sun[days .hh e:m
08 R: [mh mh
09 C: okay. and will thei ehm times change at all for the christmas period?
                    [or it stays the same?
10 R: [eh: no:
11 yeah we're closed over e:m i'll just just double check right now hang on and i'll tell you when we're [( ) closed
12 C: [okay thanks
13 (17)
14 R: we close on the twenty thi:rd?
15 C: right
16 R: and we open again on- on monday the fith.
17 (0.5)
18 C: on the fifth. [okay then. lovely.
19 R: [monday the fifth
20 C: ok[ay then thanks a lot then.
21 R: [alright then?
22 C: b[ye?
23 R: [by:e

GDGB 18 *POR50M1 *POR TRAVEL AGENCY
01 C: ((telephone rings))
02 R: good runners, good afternoon [this is]( )?
03 C: [.hhh]
04 ((noise))
05 C: >oh hallo there.< ahm (.) just one thing. can i book a ryanair flight in your agency? at all
06 R: ryana:ir?=
07 C: = yeah >do you do< with ry-ryanair.
08 R: well they've direct service (their) ticket-desk. you (.) do it over the internet eh with the credit ca:rd(s).=
09 C: = right ok .
10 R: i mean we can do it for you. using ou:r system. but it means we're charging you=
11 C: okay

12 R: =( ) twenty five pounds a ticket just for doing it
13 C: okay.-okay.(.) so it's better on the internet.
14 R: well, it's better <u>for you:</u>.
15 C: yeah
16 R: you get a better <u>fare</u>
17 C: .hh okay. well, thanks for your honesty. thank you.
18 R: ((laughs)) ahh. okay?
19 C: okay. thanks a lot. by:e?
20 R: °by:e°.

GDGB 13 *POR50M1 *POR CAR ACCESSORIES SHOP
01 C: ((telephone rings))
02 R: >(hallo) car engines?<
03 C: .hhh eh good(.)afternoon i've-v-got a smashed window on a citroen bx .hh >i wonder if i could just buy< a piece of glass to p:ut in it (ready) made?
04 R: e:hm yeah. what type of bx is it? sorry i forgot?
05 C: e:::m nineteen o heitch reg i can't remember [heitch reg
06 R: [(°   °) can you tell me what side is it?
07 C: a: driver's side.
08 (9.0)
09 R: it's not what i keep on the shelf unfortunately?
10 C: a okay
11 (1.2)
12 R: okay i could a: arrange for [sort of=
13 C: [.hhhh
14 R: =tomorrow? afternoon?
15 C: how much would it be do you know?
16 R: e:: i would say about eight pounds
17 C: eight pounds. .hh let me think about it i might just go to a glass shop and get them to cut a bit of glass to stick in it
18 R: okay?
19 C: yeah. okay thanks a lot. [by:e
20 R: [°bye°

SNGB 09 *IR24F1 *POR BOOK SHOP
01 C: ((telephone rings))
02 R: hello vivian books?
03 C: .hh hello i was wondering do you have a book by an author called grazia deledda
04 (1.0)
05 R: sorry, what was the author's name?
06 C: deledda
07 (0.8)
08 R: can you spell it
09 C: yeah d. e. l. e. d. d. a.
10 (8.5)
11 R: °i'll check it for you°

12 C: °okay thanks very much°
13 (3.5)
14 R: (we) don't have any:: i don't think we have any:: in stock. oh yes we might have one called liars fo-for love.
15 C: okay you don't have one that's called <canne al vento>, (0.2) no
16 R: °i'll check that° (.) a: a:: no [we don't
17 C: [no you don't=
18 R: =we could order it for you
19 C: okay no that it's okay. do you have any boo- any other book by italian authors? (1.2) do you know these (   )
20 R: well i'd say we have this we have another book by that author but that's=
21 C: a okay
22 R: a liars for (   ). e::m in-in translation. in english (   )
23 C: either in english or in italian.
24 R: yeah. we ha-we have got a bit in english
25 C: ah okay
26 R: a::: things like the leppe:rd by lampedusa e:m i remember (jusac) that e:m (3.5) what's her name
27 C: (   ) the book?
28 R: (   )
29 C: ah okay
30 R: (called)
31 C: ah all right, okay
32 R: things like we've got mario puzo
33 C: ah ah
34 R: yeah (there is actually quite a bit)
35 C: okay that sounds that great. can you tell me exactly where are you an i'll drop in.
36 R: yes we're e:m (.) on the corner of (fulham) road and (   ) road
37 C: e:m
38 R: see do you know (0.5) do you know (poppers green tube) station?
39 C: yeah. i do.
40 R: e:m if you come out of there and turn onto beacons field walk
41 C: hmhm
42 R: e:m and at the end turn right, we're at the bottom of that road
43 C: ah okay that sounds great thanks very much for your help
44 R: (tha:nks)
45 C: okay thanks a lot >bye bye<
46 R: bye-bye
47 C: bye.

CVGB 23 *LON33F1 *LCS FLORIST
01 C: ((telephone rings))
02 R: good afternoon? charter square florists?
03 C: .h oh hello i was just wondering ? do you do ahm deliveries?
04 R: we do
05 C: octo? okay great is that part of interflora chain or are you separate to that?
06 R: we- we're a teleflora (   )

07 C: oh: i see okay ehm .hh can you give me just some idea of the sort of price range eh just a: sort of mh: normal general purpose bouquet
08 R: where are you wanting to send [it to
09 C: [.hh be be local be within lancaster
10 R: okay well a: we start at eighteen?
11 C: mh mh
12 R: and there is two pound fifty delivery char[ge depending where it is
13 C: [right
14 R: if it's [further out
15 C: [yeah
16 R: then it gets to be a little bit more   [( )
17 C: [yeah? o::kay okay that's great that gives me a general idea. then right i'll get back to you when i've decided : e::m exactly what i want then ((slightly laughing))
18 R: okay?
19 C: okay then thanks for your help? [bye
20 R: [by:e

SNGB 04 *IR24F1 *LO BEAUTY CENTRE
01 C: ((telephone rings))
02 R: (an absolute look) how may i help ?
03 C: .hh hello i was wondering could you tell me how much it costs fo:r a manicure please
    (0.3)
04 R: a manicure?
05 C: mh[m
06 R: [fifteen pounds
07 C: okay >and do< i have to make an appointment
08 R: (we wont make an appointment) but if you come in when i'm busy (then i will fit you in)
09 C: .h ah okay that's great thanks very much?
10 R: you're welcome
11 C: bye bye
12 R: bye

SNGB 07 *IR24F1 *POR WINES
01 C: ((telephone rings))
02 R: good afternoon wine spot (   ) cristoph speaking
03 C: .hh hello i was wondering could you tell me do you stock any sardinian wines
    (.)
04 R: sorry?
05 C: do you stock any sardinian wines
    (0.6)
06 R: >what is that i don't know<
    (.)
07 C: e:m wine from sardinia
    (1.2)

152 Appendix

```
08  R: wine from
    (.)
09  C: sardinia (.) in italy
10  R: no
11  C: no? [do you know anywhere=
12  R:          [no i don't have it
13  C: =i could get it from?
14  R: oh i'm new in the business? i can't help you i'm sorry
15  C: ah okay that's no problem thanks very much
16  R: sorry bye
17  C: bhm bye bye >no problem<
```

German calls

CVD 4 *K22F *K Blumen2
```
01  C: ((Telefon klingelt))
02  R: blumen meier guten tag?
03  C: guten tag julia schnibben (.) ehm ich wollte fragen? ob sie mir- ehm etwas
    über die pflege von bonsaibäumen sagen können.
04  R: bonsai führen wir gar nicht. da gibt's in der innenstadt? sind sie von
    kö-öln?
05  C: ja
06  R: am neumarkt ist ein geschäft
07  C: mm?
08  R: (ich weiß nicht wie der heißt      ) oder so. kennen sie der haus am
    neumarkt?
09  C: ja schon
10  R: da wurde ich da mal nachfragen die haben ganz viele baumsee und
    verkaufen völlig nicht also (  ) platz (   ) das haben wir nich
11  C: m: ja gut
12  R: ja? wenn s- e:: sie da in der nähe sind ( ) wenn sie von rudolf platz
    kommen (.) e::m ( ) um neumarkt (      )
13  C: ja
14  R: (          ) aber bevor sie da überhaupt (   ) auf der rechten seite schon
15  C: da ist das geschäft
16  R: [da findet es=
17  C: [ja gut
18  R: =ich weiß nicht genau wie der heißt (.) (svelto) oder ähnlich, also? .hh da
    ist auf jeden fall ein geschäft
19  C: mm
20  R: von rudolf platz  kommen (.) e:m rechts genau auf (    ) die wo neumarkt
    platz ist
21  C: e:
22  R: ja?
23  C: vielen (dank      )
24  ((both laugh))
25  R: ok
26  C: ((continues laughing)) danke
```

27 R: danke
28 C: [tschü:ss
29 R: [tschüss

CVD 14 *K50F *K Obst und Gemüse
((Lieferant für Obst und Gemüse))
01 C: ((Telefon klingelt)
02 R: 'guten morgen?
03 C: altmann guten morgen. ich rufe an von studentenheim junkersdorf,
04 (.)
05 R: jawoll
06 C: ich wollte auch die bestellung durchgeben für morgen (0.2) und zwar hätten wir gerne (2) einmal kiwi,
07 R: einmal kiwi
08 C: eine halbe kiste bananen
09 R: mhhm
10 C: eine kiste granny smith
11 R: mh ja::
12 C: zehn kilo wirsing geschnitten
13 (2.2)
14 R: >hm<
15 C: einmal kartoffeln,
16 R: ja:
17 C: einmal champignon
18 R: mh::
19 C: einmal petersilie
20 R: mh::
21 C: und zwei kilo sprossen
22 R: mh::
23 C: ja ich denk ah vielleicht noch eine kiste tomaten >wenn es geht.< ja?
24 R: mh::
25 C: gut das wäre dann alles
26 R: salat wollen sie nicht
27 C: ne: salat haben wir noch
28 (.)
29 R: jawohl.
30 C: ja danke schön
31 R: danke::
32 C: wiederhören
33 R: wiederhören

ATD 05 *FFM30F *FFM TAGESBISTRO
01 C: ((Telefon klingelt))
02 R: mittagtisch thiel katy grimm??
03 C: ja >schönen guten tag mein name ist astrid huber (.) ich habe eine frage< (.) und zwar würde ich ganz gern für einen geburtstag einen korb herrichten lassen?
04 R: hmhm

05 C: u:nd äh >hab ich da auch schon ne bestimmte vorstellung wenn ich da nen foto mitbringen würde können sie das dann ungefähr nachmachen?
06 R: wenn wir das haben=
07 C: =ja
08 R: (.) können wir das sicherlich tun (.)
09   für wann ist das?
10 C: e:m das wäre für nächste woche ende nächste woche
11 R: ja kommen sie einfach mal vorbei und dann sprechen wir darüber
12 C: ja. .hh e:m jetzt hab ich noch eine frage? ehm da kann man alle frei machen also (käse wu::rst und-und besondere delikatessen sein? und so wEiter das geht ja alles
13 R: ja je nachdem was ( ) an
14 C: ja. e:m da ist ein besondere wein dabei gewesen wurden (   ) ich muss noch mal genau gucken .hh den-denn wir es nichts da hätten kann ich den da sie bestellen? .hh hh. also ich wurde eine kiste nehmen.
15 R: sie mussten dann ein bisschen warten dass wir mit den lahndehngeschäft besprechen dann kann ich ihnen da sagen ja oder nein
16 C: ja. (.) wann haben sie denn offen
17 R: von zehn bis zwanzig uhr
18 C: ja. okay das heißt (.) ab um zwanzig uhr is ja super. aber morgen ja wahrscheinlich nicht, ne?
19 R: morgen bis sechzehn uhr
20 C: morgen bis sechzehn uhr ja dann ich dann mal dienstag vorbei
21 R: ah ja
22 C: gut? (.) [mit=
23 R: [mm
24 C: dem habe ich jetzt gesprochen?
25 R: mit der trapet
26 C: mit der trapet ok denn ich mich einfach (angehe)
27 R: (   ) bis da es ist immer [geschäft da
28 C: [(   )
29 C: alles klar?
30 R: okey
31 C: oke danke [tschü::
32 R: [tschüß

ATD 14 *FFM22M *FFM MÖBELLADEN
01 C: ((Telefon klingelt))
02 R: (   ) guten tag?
03 C: guten tag hier ist gerhard dietrich ich hab eine frage ich hab heute ein geschenk für meine schwiegermutter nächste woche und (.) ich weiss dass sie so für fünfziger sechziger jahre sachen schwärmt=
04 R: =[hmhm]
05 C: [u:nd wollte fragen haben sie so so nierentisch das war doch damals [so
06 R: [ja ja ne: habe ich nicht .hhh ((si soffia il naso)) wer konnte das haben? .hhh ALSO DER EINZIGE LADEN der sehr gute sachen hat? ist der zueglich? .hhh e::m ist den sachsenhausen in de::r (1) weil straße. gibt es eine kleinen laden. (0.2) der so speziell hm sachen mal (   ) hat
07 C: ach so. e::m haben sie denn (.) die a- e::m telefon nummer oder?

```
08  R: ne ne ne ne ich kenne es (        )
09  C: das ist in welcher straße?
10  R: weil straße in sachsenhausen
11  C: gut. dann schaue ich mal in (   [   )
12  R:                              [bitte gerne
13  C: ja. okey? vielen [dank wieder hören
14  R:                  [tschüss
```

ATD 06 *FFM22M *FFM SCHUHGESCHÄFT
```
01  C: ((Telefon Klingelt))
02  R: von scheerbart?
03  C: .hh guten tag hier ist gerhard dietrich ähm ich (.) folgedes problem ich
    hab ähm ich möchte eine wanderung machen und zwar (.) in griechenland
    und ich bräuchte da wanderschuhe dafür (.) ham sie sowas auch?
04  R: für männer hab ich wir gar keine schuhe (.) nur für frauen
05  C: ach? sie haben nur frauen schuhe?
06  R: ja:
07  (.)
08  C: achso. (okey dann) weiss bescheid
09  R: gu[t
10  C:   [danke schön? (0.2) wieder hören
```

ATD 15 *FFM22M *FFM SCHULE VON MUSIK
```
01  C: ((Telefon klingelt))
02  R: hier hochschule für musik und darstellende kunst guten morgen?
03  C: guten morgen hier ist gerhard dietrich ähm ich wollte fragen
04  ob die frau lose schon im haus ist
05  R: .hh frau lo:se e: kontrabass
06  C: ja genau
07  R: nein noch nicht
08  C: ie- die ist immer im acht vier hundert drei
09  R: ja: aber sie ist noch nicht gekommen
10  C: ist noch nicht gekommen
11  R: ne
12  C: ja gut dann. (.) denn kann ich kann ich noten hinterlegen an der
    porter [(falls    )
13  R:    [ja selbstverständlich können sie [das machen
14  C:                                      [okey ich bin gleich da.
15  R: okey dan[ke
16  C:        [ja bis [dann
17  R:                [tschü[:ss
18  C:                     [tschüss
```

ATD 18 *FFM23F *FFM FRISEUR
```
01  C: ((Telefon klingelt))
02  R: haarstudio franz ritter guten tag?
03  C: e: guten tag ich hätte gerne gewusst wie teuer ein haarschnitt bei ihnen
    ist
04  R: für eine dame?
```

156 *Appendix*

```
05  C: >ja für mich<
06  R: ja dreiundvierzig euro
07  (0.2)
08  C: wie viel bitte
09  R: dreiundvierzig euro
10  C: a: achso .h e::m muss ich eigentlich ein termin nehmen [oder (    )
11  R:                                                        [ja ja ne
12  C: e:::m kann ich eventuell später noch mal anrufen dann schaue ich [genau
    wenn ich genau:
13  R:                                                                  [ja
14  C: = zeit habe?
15  R: mh mh gut
16  C: okey danke [schön bis- tschüss
17  R:            [tschüss
```

```
CVD 12 *K22F *K Friseur 1
01  C: ((Telefon klingelt))
02  R: >hier friseur salon scheler patrizia am apparat?<
03  C: guten tag ähm ich wollte nachfragen ob sie modelle nehmen für
    dauerwelle
04  R: ja >nehmen wir<
05  C: ah. und wann kann man? also an welchen tagen ist das?
06  R: .hh uhm das wurde gehen jeden zweiten dienstag und jeder zweiten
    donnerstag
07  C: und wie teuer ist das denn
08  (1.5)
09  R: u::m (1.2) dauerwelle normal .hh das (reinish berate klar ber) ehm würde
    sie zweiundsiebzig kosten und ä:m die modelle bezahlen bei uns die hälfte
    also >wäre das denn<
10  C: ein und dreißig
11  (1.5)
12  R: ja sechs und dreißig wäre das denn
13  C: eh: ah ja. (.) ehm (.) und dann muss man sich bei ihnen ein termin
    (holen).
14  R: .h genau
15  C: mm .h ah ja gut. vielen dank
16  R: bitte schön
17  C: wieder hören
18  R: wieder hören
```

Italian calls

```
SNI 16 *SS FARMACIA
((R is a child))
01  C: ((telephone rings))
02  R: >°pronto farmacia parenti°<
03  C: .hh e buongiorno mi scusi a che ora aprite: questo pomeriggio?
04  ((voci in sottofondo))
```

05 R: quattro e mezza
06 C: okey grazie arriveder[ci
07 R: [°>arrivederci<°

GDI 14 *NU31M2 *NU MUSEO
01 C: ((squillo))
02 R: pronto casa bianchi.
03 C: buongiorno senta chiamo per avere un'informazione .hh volevo sapere se effettuate visite guidate per per studenti di scolaresche anche al museo.
04 R: sì sì certo.
05 C: ah va benissimo. la ringrazio
06 R: prego arrive[derci
07 C: [arrivederci buona giornata

CVI 14 *SS48F1 *SS caffè
01 C: ((telephone rings))
02 R: pronto?
03 C: pronto buongiorno vorrei avere un'informazione. avete del panettone all'ananas?
04 (0.2)
05 R: un attimo che chiedo.
06 C: grazie
07 ((si sente di sottofondo chiedere se ce l'hanno))
08 R: no °mi dispiace°
09 C: non ce l'avete? .hh la ringra[zio buongiorno
10 R: [prego buongiorno

GDI 12 *NU40F1 *NU MUSEO
01 C: ((telephone rings))
02 R: pronto casa bianchi.
03 C: eh:: buonasera (è) il museo?
04 R: sì?
05 C: eh:: ascolti volevo sapere se la domenica è aperto.
06 R: sì sì è aperto tutti i giorni (.) [eh::=
07 C: [ah?
08 R: =eccetto il martedì
09 C: e ascolti mi può dire gli orari
10 R: sì allora aperto dalle dieci alle tredici
11 C: sì
12 R: e dalle sedici e trenta alle diciannove e trenta
13 C: va bene la ringrazio
14 R: di niente
15 C: arrived[erci
16 R: [arrivederci

ATI 14 *SS23F3 *SS ANTIQUARIATO
01 C: ((telephone rings))
02 R: pronto?
03 C: pronto buongiorno

158 Appendix

```
04  R: buon[gio-
05  C:       [asco:lti un'informazione io sto cercando per un regalo >un servizio
             di bicchieri antico<
06  R: e:h
07  C: ne av[ete?
08  R:      [antico no di bicchieri no .hh ho u:n bel servizio: di: >tazzine da
             caffè:<=
09  C: =ah ecco=
10  R: =e:h insomma
11  C: sono particolari
12  R: beh è bavaria? >molto buona come qualità< poi è dorato con una <pietra
       rossa: sopra> (0.2) è da dodici (.) ci sono tre caffettie:re .hh e una lattiera
13  C: quanto vengo a spendere
14  R: e dunque quello costa sulle ottocento mila lire completo. da dodici
       completo eh? bellissimo ( ) bavaria è ottimo come hh. .hh qualità
15  C: cer[to
16  R:    [non è molto antico comunque insomma è anni:: cinquanta
17  C: va bene la ringrazio allora
18  R: pre[go
19  C:    [magari passo
20  R: d'ac[cordo
21  C:     [arrivederci buongiorno

SNI 05 *SS CVS
01  C: ((telephone rings))
02  R: >sì pronto?<
03  C: hh e pronto buonasera è il centro soggiorni studi?
04  R: sì buonasera mi di[ca
05  C                    [.hh e: buonasera ascolti ho saputo che fate corsi di
       inglese: e spagnolo .hh[h
06  R:                        [sì
07  C: ecco volevo avere alcune: informazioni.
08  R: sì
09  C: verso: -magari mm. che ora posso venire? .hh m: con chi posso parlare
       ecco?
10  R: ma ss- guardi >se vuol venir siamo benissimo qua< sicuramente fino alle
       otto almeno. [quindi=
11  C:              [ah
12  R: =se vuole venire? se vuole fare un salto? ci sono io? sono: io mi chiamo
       rosario, sono: responsabi[le per quanto riguarda il settore=
13  C:                          [ah?
14  R: =corsi di lingua e viaggi studio
15  C: ho capito
16  R: quindi? se vuole venire anche adesso: sono qua? [insomma.
17  C:                                                 [va bene. magari
       mi avvicino    [stasera o:::
18  R:                [sì sa dove siamo a::l via la marmora? presso il centro
       commerciale >la marmora<. dove c'è l'arco praticamente.
19  C: ahah ho capito. va [bene
```

20 R: [d'accordo?
21 C: grazie
22 R: oppure domani mattina dopo le undici e mezza
23 C: sì?
24 R: ci sono anche all'una e un quarto. sicuramente ci sono. ma? se può venire stasera? forse è meglio così cominciamo a vedere qualcosa anche perché .hh stiamo ultimando le iscrizioni per il (.) semestre adesso
25 C: ah
26 R: e quindi: [occorrerebbe
27 C: [perché adesso inizia un nuovo corso
28 R: sì. praticamente: stiamo facendo gli ultimi inserimenti adesso .hh
29 C: ahah
30 R: quindi:: insomma, ecco
31 C: e-ecco pri:ma:: vengo meglio è insomma.
32 R: forse sì. è meglio.
33 C: va [bene=
34 R: [d'accordo?
35 C: =la ringrazio
36 R: grazie a lei ar[rivederci
37 C: [arrivederci

SNI 22 *SS DENTISTA
01 C: ((telephone rings))
02 R: >studio mirella buongiorno<
03 C: e: buongiorno sono: la mamma di chiara nietti volevo prendere un appuntamento per cortesia
04 R: cosa deve fare signora
05 C: no allora è in (0.2) in cura con il dottor mirella perché c'ha:: insomma l'apparecchietto. .hhh l'ha vista l'ultima volta il due agosto a::m ad oschiri (0.2) e quindi poi non le ha dato appuntamento dal giorno quindi volevo sapere quando la voleva vedere
06 R: mattina o pomeriggio
07 C: di pomeriggio perché va a scuola
08 (1.2)
09 R: e verso che ora signora? presto, tardi? me lo deve dire lei °perché io,°
10 C: eh? guardi noi alle quattro:: e mezza:: va bene ci- anche alle cinque?
11 (1.2)
12 R: nella prossima settimana?
13 C: non lo so quan[do::-
14 R: [perché questa setti^mana: ci sarebbe giovedì (1) però: alle se:i
15 C: no no alle sei [fa
16 R: [troppo [tardi?
17 C: [allenamento. sì sì no. diciamo intorno alle cinque se va bene
18 R: insomma alle cinque alle quattro e mezza, [cinque
19 C: [sì
20 R: e e: anche martedì? prossimo?
21 C: martedì prossimo

160 *Appendix*

22 R: alle cinque?
23 C: va bene
24 R: a nome?
25 C: e: chiara rietti
26 (1.5)
27 R: ascolti io ho segnato martedì ventiquattro? ore diciassette
28 C: va bene:
29 R: se ci sono problemi mi chiama
30 C: va bene [grazie
31 R: [buongiorno si[gnora
32 C: [bngiorno:

CVI 19 *NU42F1 *SS viaggi
01 C: ((telephone rings)) ((waiting music)) (13)
02 R: >centro viaggi buonasera sono adele?<
03 C: eh buonasera.(.) senta signora io avrei necessità:: di un'informazione per quanto riguarda (.) la continuità territoriale chi è che ha diritto agli sconti?
04 (.)
05 R: i residenti i nati in sardegna che risiedono fuori.
06 (0.5)
07 C: mhm.
08 R: mi dica com'è l-la: la persona? ha meno di ventisei anni?
09 (.)
10 C: no no no io poi volevo sapere se per esempio (.) in caso di: una famiglia separata per esempio
11 R: sì
12 C: e:: e:m uno dei coniuge è residente qu[i.-=
13 R: [ho capito
14 C: l'altro in un'altra regione della: sard- della:: dell'italia
15 R: sì
16 C: e: i::l minore anche se si trova in un'altra regione ha diritto allo sconto?
17 R: ma è in sardegna il minore?
18 C: no.
19 R: no allora no. sono sono i lati::: be no aspetti il minore? quanti anni ha? se ha meno di ventisei anni sì?
20 C: sì sì ha meno di-
21 R: e allora sì.
22 C: e allora ha dirit[to com-
23 R: [se è così sì
24 C: <ho capito>
25 R: perché è pe:r gli emigrati? quelli nati in sardegna? per i residenti per i giovani
26 (.)
27 C: °sì°
28 R: per i senior. cioè più di sessant'anni
29 C: °sì°
30 R: pe:r eh gli studenti universitari al di sotto dei ventisette anni?
31 (0.2)
32 C: ahah.

33 R: e comunque se è un minore?
34 (0.2)
35 C: ho [capito
36 R: [è un giovane
37 C: ho capito .hh e per quanto riguarda (.) i prezzi (.) e:m alghero roma? quanto costa.
38 R: >alghero roma e roma alghero andata e ritorno tasse incluse ottantasei virgola ventitre euro
39 (0.2)
40 C: e: invece. olbia roma?
41 (.)
42 R: olbia roma >forse viene di più< ((talking to the colleagues in the agency)) quanto costava olbia roma? novanta:: (.) euro. qualcosa del genere. cambia di poco comunque sono le tasse aeroportuali più che altro.
43 C: e per quali altre città è valido: [la sconto
44 R: [( )
45 C: solo per queste due °città°=
46 R: =c'è olbia bologna: (0.2) o::m però insomma ci sono sono diverse tariffe/ (.) nel senso che e: bologna è fino a marzo per esempio.
47 (0.8)
48 C: cioè lo sconto è fino a marzo poi non c'è più::
49 R: poi non si sa se lo rimettono.
50 C: °ah ho capito°. e quanto costa olbia: bologna?
51 R: sola andata settantuno euro. (.) settantasei euro sarebbe.
52 C: °ho capito°e::m è necessario prenotare molto tempo prima?
53 R: sì perché e: i posti sono limitati (.) almeno su olbia bologna poi dipende insomma da dove deve andare e quando deve partire
54 C: e per roma per esempio. la la alghero e:: (.) roma e::m quella linea è molto trafficata? cioè bisogna prenotare molto t-
55 R: per adesso c'è posto però e: più che altro al fine settimana ci sono problemi.
56 (.)
57 C: °a: ho capito° mentre?
58 R: e: è u:n pochino vasta la cosa
59 C: °ho capito° va bene la ringrazio tanto. buonasera
60 R: >rivederci<

ATI 03 *SS23F3 *SS PARRUCCHIERE3
01 C: ((telephone rings))
02 R: >alma buonasera<
03 C: buonasera scusi un'informazione >voi siete: mh fate sconti pe:r studenti< per tagli [e piega
04 R: [no hh
05 C: niente. non siete neanche affiliati arco per caso?
06 R: no no
07 C: ah ho capito. quanto costa comunque [per il-
08 R: [ventidue euro
09 C: ah? va bene.
10 R: ( )

162 Appendix

```
11  C: a(.)scolti allora richiamo per un appuntamento
12  R: va[ bene
13  C:    [va bene? grazie ar[rivederci
14  R:                       [arrivederci arriv°ederci°
```

GDI 19 *NU24F1 *SS OTTICA
```
01  C: ((squillo))
02  R: arte ottica?
03  C: .hh eh buonasera vorrei chiedere un'informazione .hhh eh dunque vorrei
    acquistare delle lenti a contatto mensili .hhh però ho smarrito- oh: il referto
    dell'oculista (.) hh (h) è
    po[ssibile
04  R:[e l'occhiale ce l'ha?
05  C: sì l'occhiale sì l'ho preso da voi tra l'altro=
06  R: =sì va beniss[imo.porta l'occhiale?=
07  C:             [.hhh
08  C: =ah ah?
09  R: noi leggiamo dall'occhiale::
10  C: la gradazione
11  R: sì (per prendere) la gradazione
12  C: mh h okey grazie e:?
13  R: prego
14  C: arrive[derci
15  R:      [buonasera
```

GDI 11 *NU40F1 *NU ASL
```
01  C: ((squillo))
02  R: °pronto°?
03  C: eh:: buongiorno .hh eh:: mi scusi io (.) mi serve un'informazione.
04  R: °mi dica°.
05  C: .hh eh:: allora sono una::hm una studentessa .hh eh:: devo andare in
    viaggio in inghilterra.
06  R: °sì (.) mi dica°.
07  C: vorrei sapere (.) eh::hm per un mese come p- come devo fare per
    l'assistenza sanitaria.
08  R: un mese(.)solamente?
09  C: sì per un mese .
10  R: niente:: viene qua col libretto sanitario- oh[::
11  C:                                              [sì
12  R: e noi rilasciamo il modello e centoundici.
13  C: ecco
14  R: °e nient'altro°
15  C: e de- devo compilare questo modello oppure?
16  R: no no no lo compiliamo noi? [il modello
17  C:                             [ah ecco e così io sono coperta per quel mese:
18  R: (esatto)
19  C: va bene va bene la ringrazio.
20  R: (°buongiorno°)
21  C: arrivederci
```

CVI 8 *NU42F1 *SS libreria2
01 C: ((squillo))
02 R: °libri scolastica°
03 C: e:: buonasera senta avrei necessità di un'informazione, avete dei testi giuridici?
04 R: (no signora) ma di che genere? pe:r dei concorsi?
05 C: >no no no< i codici e:: i manuali di diritto e di procedura civile e penale
06 R: no non li abbiamo
07 C: la ringrazio. buonasera

SNI 15 *SS CVS
01 C: ((squillo))
02 R: pronto?
03 C: e: pronto buongiorno potrei parlare co:l signor caravag[gio?
04 R: [sono io signora [buongiorno
05 C: [.hh buongiorno ascolti volevo sapere se organizzate anche: viaggi: di studio in spagna
06 R: sì certamente
07 C: sì:?
08 R: sì sì sì sì sì
09 C: ecco a me interessa mm più o meno il periodo di agosto
10 R: certo
11 C: e:: volevo sapere se sono ancora in tempo [oppure:: =
12 R: [sì sì
13 C: =è tardi
14 R: be? ci mancherebbe [noi andiamo avanti=
15 C: [mh
16 R: =fino ad agosto poi con le prenotazioni. (.)
17 C: ho ca[pito
18 R: [praticamente::=
19 R: =è a ciclo continuo. ((laughter)) chiaramente, il discorso è questo,
20 C: sì
21 R: prima si prenota? e meglio è? perché. perché così riusciamo a trovare? (.) anche più disponibilità per quanto riguarda l'alloggio
22 C: ho capito
23 R: e:: poi soprattutto in spagna c'è il problema anche del posto in aereo. dei voli?
24 C: mh mh
25 R: e::::: che non è facile indubbiamente trovare posto in spagna
26 C: sì
27 R: anche perché? per l'inghilterra adesso (.) un po' abbiamo la possibilità cioè grazie a dio abbiamo la possibilità con la ryan air?
28 C: mh mh
29 R: di [trovare delle tariffe convenienti,
30 C: [certo
31 R: con la spagna? è un po' più difficile. dove le interesserebbe in spagna?
32 C: ma e:: non ho un e:: diciamo una meta precisa. io? (.)studio lingue

33 [e::
34 R: [>ho capito. ha visto la locandi-< ha saputo:: da: tramite:: ha saputo di noi tramite qualcuno? che [è partito con noi
35 C: [sì tramite miei colleghi che::
36 R: ah
37 C: sono partiti con voi sì
38 R: ((caugh)) chiedo scusa
39 C: e: niente
40 R: ((continues caughing))
41 C: e quindi: appunto. la mia: non ho una destinazione precisa   por- mi interessa appunto trascorrere un periodo lì e::m magari frequentare una scuola
42 R: be certo certo
43 C: e:: quindi avere insomma? non so? andare in famiglia?
44 R: certamente? guarda, tanto ci possiamo dare del tu visto che
45 tu se[i:: >sicuramente< più giovane di me=
46 C: [((laughter)) e:: sì sì
47 R: io ne ho trentatre tu quanti ne hai scusami?
48 C: e: ventidue
49 R: e:: allora dai [dammi del tu (   )
50 C: [e: sì ((embarrassed))
51 R: senti:: come ti chiami mi hai detto?
52 C: e:: sandra
53 R: sandra. allora guarda? per la spagna abbiamo tutto, ma proprio tutto tutte le destinazioni possibili e immaginabili (.) allora barcellona   e::m madrid e:
54 C: mh mh
55 R: sivi- e valenzia siviglia e:: d- salamanca
56 C: sì
57 R: malaga e::m mar[bella addirittura marbella
58 C: [ho capi-
59 R: e:: guar[da? (ti dico) granada
60 C: [di tutto insomma
61 R: e:: qualsiasi destinazione tu voglia noi ce l'abbiamo [per quanto riguarda la spagna
62 C: [ho capito
63 R: e:: lavoriamo con la don quichotte
64 C: sì
65 R: poi con l'extension language centre a ma- ma- marbella
66 C: mh mh
67 R: tutte destinazioni (.) belle?
68 C: ho capi[to
69 R: [e le scuole sono: ottime tutte visit-visi[tate da me
70 C: [certo sì sì
71 R: per cui-
72 C: me ne hanno parlato bene mh.
73 R: sì no e ti ringrazio
74 C: [((laughs))
75 R: [>son ringraziamenti personali< chi è che che conosci che è partito >se ti posso chiedere< scusami?

```
76   C: e: no i:n in spagna no? [e::
77   R:                          [eh?
78   C: però in altre:: insomma con- per altre desti[nazioni:: in inghilterra
79   R:                                              [e chi e chi cono- chi ti
     ha parlato di noi se ti posso chiedere
80   C: e:: mi ha parlato:: una certa paola:: (0.2) cuc[cin- eh
81   R:                                                [ah? sì sì sì cer[to
82   C:          [è partita l'anno scorso [con voi?
83   R:                                   [e partirà anche quest'anno con noi.
     [va a brema
84   C:   [in germania esatto sì
85   R: e in germania, certo?
86   C: ah ah
87   R: ascolta? [e: se tu hai=
88   C:         [sì
89   R: =la possibilità di fare un salto?    [da noi
90   C:                                      [certo? magari mi avvicino per avere
     ulteriori informa[zioni
91   R:                [ecco. anche stamattina se vuoi? verso: fine mattina:[ta
     quando vuoi tu=
92   C:               [mh eh?
93   R: =quando preferisci
94   C: stamattina non so se posso. magari domani mattina mi avvicino=
95   R: ecco? io ci sono dopo le <undici> e mezzo >la mattina mi trovi
     sicuramen[te<? e la sera=
96   C:       [ho capito
97   R: =dopo le cinque e un quarto cinque e mezzo anche fino a tardi.
98   C: mh mh?
99   R: la mattina? ma anche fino all'una e mezza almeno ci sono sicuramente
100  C: va bene
101  R: d'accordo?
102  C: okey ti ringra[zio
103  R:               [bene >sai dove siamo allora< sì?
104  C: sì sì [mso::
105  R:      [su al >centro commerciale luna e sole<
106  C: sì lo so. mh. gra[zie sì
107  R:                  [d'accordo, cia:[o
108  C:                                  [arrivederci
```

References

Aston, G. (ed.) 1988a. *Negotiating Service, Studies in the Discourse of Bookshop Encounters*. The PIXI project. Bologna, CLUEB.
Aston, G. 1988b. What's a Public Service Encounter Anyway? In G. Aston, *Negotiating Service, Studies in the Discourse of Bookshop Encounters*. The PIXI project. Bologna, CLUEB, 25–42.
Atkinson, J.M. and Drew, P. 1979. *Order in Court: the Organization of Verbal Interaction in Judicial Settings*. London, Macmillan.
Atkinson, J.M. and Heritage, J. 1984. *Structures of Social Action: Studies in Conversation Analysis*. Cambridge, Cambridge University Press.
Auer, P. 2005. Projection in Interaction and Projection in Grammar. *Text*, 25/1, 7–36.
Barnes, S. and Armstrong, E. 2010. Conversation after Right Hemisphere Brain Damage: Motivations for Applying Conversation Analysis. *Clinical Linguistics and Phonetics*, 24/1, 55–69.
Beach, W. 1993. Transitional Regularities for 'Casual' 'Okay' Usages. *Journal of Pragmatics*, 19, 325–52.
Bercelli, F. and Pallotti, G. 2002. Conversazioni telefoniche. In C. Bazzanella (ed.), *Contesti e forme di interazione verbale*. Milan, Guerini Studio, 177–93.
Bergman, M. and Kasper, G. 1993. Perception and Performance in Native and Nonnative Apology. In G. Kasper and S. Blum-Kulka (eds), *Interlanguage Pragmatics*. New York, Oxford University Press, 82–107.
Bergmann, J.R. 1993. Alarmiertes Verstehen: Kommunikation in Feuerwehrnotrufen. In T. Jung and S. Muller-Doohm (eds), *Wirklichkeit im Deutungsprozess: Verstehen und Methoden in der Kultur- und Sozialwissenschaften*. Frankfurt/M., Suhrkamp, 283–328.
Billig, M. 1999. Conversation Analysis and the Claims of Naivety. *Discourse in Society*, 10/4, 572–6.
Bilmes, J. 1988. The Concept of Preference in Conversation Analysis. *Language in Society*, 17, 161–81.
Bonu, B. 1995. Questions sur la préférence en Analyse de Conversation: hiérarchisation dans les entretiens de recrutement. *Cahiers de l'ILSL*, 7, 199–230.
Bowles, H. and Pallotti, G. 2004. Conversation Analysis of Opening Sequences of Telephone Calls to Bookstores in English and Italian. *Textus*, 17/1, 63–88.
Boyle, R. 2000. Whatever Happened to Preference Organisation? *Journal of Pragmatics*, 32, 583–604.
Brodine, R. 1991. Requesting and Responding in Italian and English Service Encounters. In *Dialoganalyse III, Referate der 3. Arbeitstagung Bologna 1990*. Tübingen, Max Niemeyer, 293–305.
Cameron, D. 2000. *Good to Talk?* London, Sage.
Church, A. 2009. *Preference Organisation and Peer Disputes. How Young Children Resolve Conflict*. Farnham, Ashgate Publishing Limited.
Clark, H. 1994. Discourse in Production. In M. A. Gernsbacher (ed.), *Handbook of Psycholinguistics*. San Diego, Academic Press, 985–1021.

Colamussi, A. and Pallotti, G. 2003. Le aperture di telefonate in italiano e spagnolo. In E.M. Thüne and S. Leonardi (eds), *Telefonare in diverse lingue*. Milan, Franco Angeli, 91–111.
Couper-Kuhlen, E. and Selting, M. 1996. *Prosody in Conversation: Interactional Studies*. Cambridge, Cambridge University Press.
De Stefani, E. 2006. Le chiusure conversazionali nell'interazione al banco di un supermercato. In Y. Bürki and E. De Stefani, *Trascrivere la lingua. Dalla filologia all'analisi conversazionale*. Berne, Peter Lang, 369–403.
De Stefani, E. 2008. Diciamo dieci ... o nove e mezzo. La manipolazione degli oggetti nelle sequenze di quantificazione. In M. Pettorino, A. Giannini, M. Vallone and R. Savy (eds), *La comunicazione parlata*: Vol. 2. *Atti del congresso di Napoli*, 23–25/2/2006, Naples, Liguori, 1059–80.
Drew, P. and Heritage, J. (eds) 1992. *Talk at Work*. Cambridge, University Press.
Drew, P. and Heritage, J. 2006. *Conversation Analysis*. London, Sage.
Drew, P. 2005. Foreword: Applied Linguistics and Conversation Analysis. In K. Richards and P. Seedhouse (eds), *Applying Conversation Analysis*. Basingstoke, Palgrave Macmillan, xvi–xx.
Drummond, K. and Hopper, R. 1993. Some Uses of Yeah. *Research on Language and Social Interaction*, 26/2, 203–12.
Du Bois, J., Schuetze-Coburn, S., Paolino D. and Cumming, S. 1993. Outline of Discourse Transcription. In J.A. Edwards and M.D. Lampert (eds), *Talking Data: Transcription and Coding Methods for Language Research*. Hillsdale, NJ, Lawrence Erlbaum, 45–89.
Eggins, S. and Slade, D. 1997. *Analysing Casual Conversation*. London, Cassel.
Finch, L. C. 2001. *L'addetto al call center. La voce dell'azienda*. Milan, Franco Angeli.
Ford, C. E., Fox, B. A. and Thompson, S. A. 1996. Practices in the Construction of Turns: the 'TCU' Revisited. *Pragmatics*, 6/3, 427–54.
Ford, C. E., Fox, B. A. and Thompson, S. A. 2002a. *The Language of Turn and Sequence*. Oxford, Oxford University Press.
Ford, C. E., Fox, B. A. and Thompson, S. A. 2002b. Social Interaction and Grammar. In M. Tomasello (ed.), *The New Psychology of Language: Cognitive and Functional Approaches to Language Structure*, Vol. 2. Mahwah, NJ, Lawrence Erlbaum, 119–43.
Ford, C. E. and Thompson, S. 1996. Interactional Units in Conversation: Syntactic, Intonational, and Pragmatic Resources for the Management of Turns. In E. Ochs, E. A. Schegloff and S. A. Thompson (eds), *Interaction and Grammar*. Cambridge, Cambridge University Press, 134–84.
Ford, C. E. and Wagner, J. 1996. Interaction-Based Studies of Language. *Special Issue of Pragmatics*, 6/3, 277–456.
Fox, B.A. 2007. Principles Shaping Grammatical Practices: an Exploration. *Discourse Studies*, 9, 299–318.
Fox, B.A. and Thompson, S. 2010. Responses to Wh-Questions in English Conversation. *Research on Language and Social Interaction*, 43/2, 133–56.
Gardner, R. 1997. The Conversation Object Mm: a Weak and Variable Acknowledging Token. *Research on Language and Social Interaction*, 30/2, 131–56.
Gavioli, L. 1995. Turn Initial versus Turn Final Laughter: Two Techniques for Initiating Remedy in English/Italian Bookshop Service Encounters. *Discourse Processes*, 19, 369–84.

Gavioli, L. 1999. Alcuni meccanismi di base nell'analisi della conversazione. In R. Galatolo and G. Pallotti (eds), *La conversazione*. Milan, Cortina, 43–65.
Gilbert, G.N. and Mulkay, M. 1984. *Opening Pandora's Box*. Cambridge, Cambridge University Press.
Goffman, E. 1981. *Forms of Talk*. Philadelphia, University of Pennsylvania.
Goi, A. 2005. *Lavorare al call center. Manuale di formazione e autoformazione*. Milan, Franco Angeli.
Goodwin, C. 1979. The Interactive Construction of a Sentence in Natural Conversation. In G. Psathas (ed.), *Everyday Language: Studies in Ethnomethodology*. New York, Irvington, 97–121.
Goodwin, C. 1981. *Conversational Organization: Interaction between Speakers and Hearers*. New York, Academic Press.
Greatbatch, D. 1988. A Turn-Taking System for British News Interviews. *Language in Society*, 17, 401–30.
Gumperz, J. 1982. *Discourse Strategies*. Cambridge, Cambridge University Press.
Guthrie A. 1997. On the Systematic Deployment of Okay and Mmhmm in Academic Advising Sessions. *Pragmatics*, 7, 397–415.
Halmari, H. 1993. Intercultural Business Telephone Conversations: a Case of Finns vs. Anglo-Americans. *Applied Linguistics*, 14, 408–30.
Hart, C. W. L., Heskett, J. L. and Sasser, W. E. 1990. The Profitable Art of Service Recovery. *Harvard Business Review*, July, 148–54.
Heritage, J. 1984a. A Change-of-State Token and Aspects of its Sequential Placement. In J.M. Atkinson and J. Heritage (eds), *Structures of Social Action: Studies in Conversational Analysis*. Cambridge, Cambridge University Press, 299–345.
Heritage, J. 1984b. *Garfinkel and Ethnomethodology*. Cambridge, Polity Press, 265–80.
Heritage, J. 1998. Oh-prefaced Responses to Inquiry. *Language in Society*, 27, 291–334.
Heritage, J. 1999. Conversation Analysis at Century's End: Practices of Talk-in-Interaction, Their Distributions, and Their Outcomes. *Research on Language and Social Interaction*, 32/1–2, 69–76.
Heritage, J. 2002. Oh-prefaced Responses to Assessments. A Method of Modifying Agreement/disagreement. In C. Ford, B. Fox and S.A. Thompson (eds), *The Language of Turn and Sequence*. Oxford, Oxford University Press, 196–224.
Heritage, J. 2008. Conversation Analysis as a Social Theory. In B. Turner (ed.), *The New Blackwell Companion to Social Theory*. Oxford, Blackwell, 300–20.
Hester, S. and Francis, D. 2001. Is Institutional Talk a Phenomenon? Reflections on Ethnomethodology and Applied Conversation Analysis. In A. McHoul and M. Rapley (eds), *How to Analyse Talk in Institutional Settings: a Casebook of Methods*. London, New York, Continuum, 206–17.
Hopper, R. 1991. Hold the Phone. In *Talk and Social Structure*. Cambridge, UK, Polity Press, 217–231.
Hopper, R. and Doany, N. 1988. Telephone Openings and Conversational Universals: a Study in Three Languages. In S. Ting-Toomey and F. Korzenny (eds), *Language, Communication and Culture*. Newbury Park, Sage, 157–79.
Hutchby, I. 2001. *Conversation and Technology, from the Telephone to the Internet*. Cambridge, UK, Polity Press.
Hutchby, I. and Wooffitt, R. 1998. *Conversation Analysis*. Cambridge, UK, Polity Press.

Hymes, D. 1972. Models of the Interaction of Language and Social Life. In J. Gumperz and D. Hymes (eds), *Directions in Sociolinguistics: the Ethnography of Communication*. New York, Holt, Rinehart and Winston, 35–71.
Jefferson, G. 1972. Side Sequences. In D.N. Sudnow (ed.), *Studies in Social Interaction*. New York, Free Press, 294–338.
Jefferson, G. 1980. On 'Trouble-Premonitory' Response to Inquiry. *Sociological Inquiry*, 50/3–4, 153–85.
Jefferson, G. 1985. Notes on a Systematic Deployment of the Acknowledgement Tokens 'Yeah' and 'Mmhm'. *Papers in Linguistics*, 17/2, 197–216.
Joyce, H., Nesbitt, C., Scheeres, H., Slade, D. and Solomon, N. 1995. *Effective Communication in the Restructured Workplace*. 2 vols. Victoria, Australia, National Food Industry Training Council.
Jucker, A. H. and Ziv, Y. 1998. *Discourse Markers. Descriptions and Theory*. Amsterdam, John Benjamins Publishing.
Lee, S. H. 2011a. Managing Nongranting of Customers' Requests in Commercial Service Encounters. *Research on Language and Social Interaction*, 44/2, 109–34.
Lee, S. H. 2011b. Responding at a Higher Level: Activity Progressivity in Calls for Service. *Journal of Pragmatics*, 43, 904–17.
Leidner, R. 1993. *Fast Food, Fast Talk: Service Talk and the Routinization of Everyday Life*. Berkeley, Calif., University of California Press.
Lerner, G. 1996. On the 'Semi-Permeable' Character of Grammatical Units in Conversation: Conditional Entry into the Turn Space of Another Speaker. In E. Ochs, E. Schegloff and S.A. Thompson (eds), *Interaction and Grammar*. Cambridge, Cambridge University Press, 238–76.
Lesser, R. 2003. When Conversation Is not Normal: the Role of Conversation Analysis in Language Pathology. In C. L. Prevignano and P. J. Thibault (eds), *Discussing Conversation Analysis: the Work of Emmanuel A. Schegloff*. Philadelphia/Amsterdam, John Benjamins Publishing Company, 141–55.
Levinson, S. 1983. *Pragmatics*. Cambridge, Cambridge University Press.
Lieflander-Koistinen, L. 1992. Asking for Information: Differences in the Interactional Structure of Openings in Finnish and German Telephone Calls. In H. Nyyssönen and L. Kuure (eds), *Acquisition of Language – Acquisition of Culture*. AFinLA, Yearbook 1992, 20–223.
Lindström, J. 2006. Grammar in the Service of Interaction: Exploring Turn Organization in Swedish. *Research on Language and Social Interaction*, 39, 81–117.
Luke, K.K. and Pavlidou, T. 2002. *Telephone Calls*. Amsterdam, John Benjamins.
McHoul, A.W. 1978. The Organization of Turns at Formal Talk in the Classroom. *Language in Society*, 7, 183–213.
Mehan, H. 1979. *Learning Lessons: Social Organization in the Classroom*. Cambridge, Mass., Harvard University Press.
Menzler-Trott, E. 1999. *Call-Center-Management: ein Leitfaden für Unternehmen zum effizienten Kundendialog*. Munich, Beck.
Mey, J. 1993. *Pragmatics: an Introduction*. London, Blackwell.
Mondada, L. 2005. L'analyse de corpus en linguistique interactionelle: de l'étude de cas singuliers à l'étude de collections. In A. Condamines (ed.), *Sémantique et corpus*. Paris, Hermes, 75–108.
Monzoni, C. and Zorzi, D. 2003. Le telefonate d'emergenza: un confronto fra l'italiano e l'inglese. In E. Thüne and S. Leonardi (eds), *Telefonare in diverse lingue*. Milan, F. Angeli, 163–81.

References

Ochs, E., Schegloff, E. and Thompson, S. (eds) 1996. *Interaction and Grammar*. Cambridge, Cambridge University Press.
Olshtain, E. 1989. Apologies across Languages. In S. Blum-Kulka, J. House and G. Kasper (eds), *Cross-Cultural Pragmatics: Requests and Apologies*. Norwood, NJ: Ablex, 155–73.
Ono, T. and Thompson, S. 1994. Unattached NPs in English Conversation. *Proceedings of the Twentieth Annual Meeting of the Berkeley Linguistics Society*. Berkeley, University of California Press, 402–19.
Ono, T. and Thompson, S. 1995. The Dynamic Nature of Conceptual Structure Building: Evidence from Conversation. In A. Goldberg (ed.), *Conceptual Structure, Discourse and Language*. Stanford: Center for the Study of Language and Information, 25–34; reprinted in E. Ventola (ed.), *The New Courant*, 4, 25–34.
Pallotti, G. 2007. Conversation Analysis: Methodology, Machinery, and Application to Specific Settings. In H. Bowles and P. Seedhouse (eds), *Conversation Analysis and Language for Specific Purposes*. Berlin, Peter Lang, 37–67.
Pallotti, G. and Varcasia, C. 2008. A Comparative Study of Service Telephone Call Openings in Five European Languages. *Journal of Intercultural Communication*, 17.
Pavlidou, T. 1994. Contrasting German–Greek Politeness and the Consequences. *Journal of Pragmatics*, 21, 487–511.
Pomerantz, A. 1984. Agreeing and Disagreeing with Assessments: Some Features of Preferred/Dispreferred Turn Shapes. In J.M. Atkinson and J. Heritage (eds), *Structures of Social Action: Studies in Conversational Analysis*. Cambridge, Cambridge University Press, 57–101.
Psathas, G. 1990. Introduction: Methodological Issues and Recent Developments in the Study of Naturally Occurring Interaction. In G. Psathas (ed.), *Interactional Competence*. Washington, DC, University Press of America, 1–30.
Psathas, G. 1995. *Conversation Analysis*. Thousand Oaks, Sage.
Ronchi, M. 1999. *Call center: istruzioni per l'uso. Un contatto diretto con gli affari*. Milan, Franco Angeli.
Sacks, H. 1987. On the Preferences for Agreement and Contiguity in Sequences in Conversation. In G. Button and J. R. E. Lee (eds), *Talk and Social Organisation*. Clevedon, England, Multilingual Matters Ltd, 54–69.
Sacks, H., Schegloff, E. and Jefferson, G. 1974. A Simplest Systematics for the Organization of Turn-Taking for Conversation. *Language*, 50, 697–735.
Sarangi, S. and Roberts, C. 1999. *Talk, Work and Institutional Order, Discourse in Medical, Mediation and Management Settings*. Berlin, New York, Mouton de Gruyter.
Schegloff, E. 1968. Sequencing in Conversational Openings. *American Anthropologist*, 70, 1075–95.
Schegloff, E. 1972. Notes on a Conversational Practice: Formulating Place. In D. Sudnow (ed.), *Studies in Social Interaction*. New York, Free Press, 75–119.
Schegloff, E. 1982. Discourse as Interactional Achievement: Some Uses of 'uh huh' and Other Things that Come between Sentences. In D. Tannen, *Georgetown University Roundtable on Languages and Linguistics, Analyzing Discourse: Text and Talk*, 71–93.
Schegloff, E. 1986. The Routine as Achievement. *Human Studies*, 9, 111–51.
Schegloff, E. 1987. Recycled Turn Beginnings: a Precise Repair Mechanism in Conversation's Turn-Taking Organisation. In G. Button and J. Lee (eds), *Talk and Social Organisation*. Clevedon, England, Multilingual Matters, 70–85.

Schegloff, E. 1988. On an Actual Virtual Servo-Mechanism for Guessing Bad News: a Single Case Conjecture. *Social Problems*, 35/4, 442–57.
Schegloff, E. 1992. In Another Context. In A. Duranti and C. Goodwin (eds), *Rethinking Context*. Cambridge, Cambridge University Press, 193–227.
Schegloff, E. 1993. Reflections on Quantification in the Study of Conversation. *Research on Language and Social Interaction*, 26, 99–128.
Schegloff, E. 1996. Turn Organization: One Intersection of Grammar and Interaction. In E. Ochs, E. Schegloff and S. Thompson, *Interaction and Grammar*. Cambridge, Cambridge University Press, 52–133.
Schegloff, E. 1999. Schegloff's Texts as Billig's Data: a Critical Reply. *Discourse in Society*, 10/4, 558–72.
Schegloff, E. 2002. Reflections on Research on Telephone Conversation: Issues of Cross Cultural Scope and Scholarly Exchange, Interactional Import and Consequences. In K.K. Luke and T.S. Pavlidou (eds), *Telephone Calls: University and Diversity in Conversational Structure across Languages and Cultures*. Amsterdam, Benjamins, 249–81.
Schegloff, E. 2007. *Sequence Organization in Interaction: a Primer in Conversation Analysis*. Cambridge, Cambridge University Press.
Schegloff, E., Jefferson, G. and Sacks, H. 1977. The Preference for Self-Correction in the Organization of Repair in Conversation. *Language*, 53/2, 361–82.
Schegloff, E., Ochs, E. and Thompson, S. 1996. Introduction. In E. Ochs, E. A. Schegloff and S. Thompson (eds), *Interaction and Grammar*. Cambridge, Cambridge University Press, 1–51.
Schegloff, E. and Sacks, H. 1973. Opening up Closings. *Semiotica*, VIII.4, 289–327.
Schiffrin, D. 1987. *Discourse Markers*. Cambridge, Cambridge University Press.
Seedhouse, P. 2005. Conversation Analysis as Research Methodology. In K. Richards and P. Seedhouse (eds), *Applying Conversation Analysis*. New York, Palgrave Macmillan, 251–66.
Selting, M. 1994. Konstruktionen am Satzrand als interaktive Ressource in natürlichen Gesprächen. In B. Haftka (ed.), *Was determiniert Wortstellungsvariation? Studien zu einem Interaktionsfeld von Grammatik, Pragmatik und Sprachtypologie*. Opladen, Westdeutscher Verlag, 299–318.
Selting, M. 1996. On the Interplay of Syntax and Prosody in the Constitution of Turn-Constructional Units and Turns in Conversation. *Pragmatics*, 6/3, 357–88.
Selting, M. 2000. The Construction of Units in Conversational Talk. *Language in Society*, 29, 477–517.
Selting, M. and Couper-Kuhlen, E. 2001. *Studies in Interactional Linguistics. Grammar and Discourse*. Amsterdam, Benjamins.
Sidnell, J. and Stivers, T. (eds) 2012. *Handbook of Conversation Analysis*. London, Blackwell.
Silverman, D. 1987. *Communication and Medical Practice: Social Relations in the Clinic*. London, Sage.
Silverman, D. 1998. *Harvey Sacks. Social Science and Conversation Analysis*. Cambridge, UK, Polity Press.
Svennevig, J. 2004. Other Repetition as Display of Hearing, Understanding and Emotional Stance. *Discourse Studies*, 6/4, 489–516.
Tannen, D. 1989. *Talking Voices: Repetition, Dialogue, and Imagery in Conversational Discourse*. Cambridge, Cambridge University Press.
ten Have, P. 1999. *Doing Conversation Analysis*. London, Sage.

ten Have, P. 2001. Applied Conversation Analysis. In A. McHoul and M. Rapley (eds), *How to Analyse Talk in Institutional Settings: a Casebook of Methods*. London, New York, Continuum, 3–11.
ten Have, P. 2002. Comparing Telephone Call Openings: Theoretical and Methodological Reflections. In K.K. Luke and T.S. Pavlidou (eds), *Telephone Calls: University and Diversity in Conversational Structure across Languages and Cultures*. Amsterdam, Benjamins, 233–48.
Thieme, K. H. and Steffen, W. 1999. *Call Center. Der professionelle Dialog mit dem Kunden*. Landsberg, Moderne Industrie Verlag.
Thüne, E. M. 2003. Telefonate di servizio in tedesco: esempi di comunicazione tra nativi e non-nativi. In E. Thüne and S. Leonardi (eds), *Telefonare in diverse lingue*. Milan, Franco Angeli, 133–62.
Thüne, E. M. and Leonardi, S. (eds) 2003. *Telefonare in diverse lingue*. Milan, Franco Angeli.
Trosborg, A. 1987. Apology Strategies in Natives/Non-Natives. *Journal of Pragmatics*, 11, 147–67.
Varcasia, C. 2003a. Chiamate di servizio in Italia e Germania: aperture a confronto. In E. Thüne and S. Leonardi (eds), *Telefonare in diverse lingue*. Milan, Franco Angeli, 112–32.
Varcasia, C. 2003b. Le conversazioni telefoniche di servizio. In G. Contu (ed.) *Annali della Facoltà di Lingue e Letterature Straniere*, vol. 1, Università di Sassari, 425–48.
Varcasia, C. 2006. Telefonate tra inglesi e italiani. La gestione delle aperture di servizio in italiano. In E. Banfi, L. Gavioli, C. Guardiano and M. Vedovelli (eds), *Atti del V Congresso dell'Associazione Italiana di Linguistica Applicata* (AItLA). Perugia, Guerra, 283–303.
Varcasia, C. 2007. English, German and Italian Responses in Telephone Service Encounters. In P. Seedhouse and H. Bowles (eds), *Conversation Analysis and Language for Specific Purposes*. Bern, Peter Lang, 217–44.
Varcasia, C. 2008. Chiamate di servizio in diverse lingue: il ruolo del chiamante. In M. Pettorino, A. Giannini, M. Vallone and R. Savy (eds), *La Comunicazione Parlata*, Atti del Congresso Internazionale di Napoli, 23–25 February 2006, Liguori, vol. II, 1272–92.
Vinkhuyzen, E. and Szymanski, M. H. 2005. Would You Like to Do It Yourself? Service Requests and Their Non-Granting Responses. In K. Richards and P. Seedhouse (eds), *Applying Conversation Analysis*. Basingstoke, Palgrave Macmillan, 91–106.
Wakin, M. and Zimmerman, D. 1999. Reduction and Specialization in Emergency and Directory Assistance Calls. *Research on Language and Social Interaction*, 32/4, 409–37.
Whalen, M.R. and Zimmerman, D. 1987. Sequential and Institutional Contexts in Calls for Help. *Social Psychology Quarterly*, 30/2, 172–85.
Wiencke, W. and Koke, D. 1999. *Call Center Praxis. Den telefonischen Kundenservice erfolgreich organisieren*. Stuttgart, Schäffer-Poeschel Verlag.
Wooffitt, R. 2005. *Conversation Analysis and Discourse Analysis: a Comparative and Critical Introduction*. London, Sage Publications.
Wootton, A. J. 1981. The Management of Grantings and Rejections by Parents in Request Sequences. *Semiotica*, 37/1–2, 59–89.
Yule, G. 1996. *Pragmatics*, Hong Kong, Oxford University Press.

Zimmerman, D. 1984. Talk and Its Occasions: the Case of Calling the Police. In Schiffrin, D. (ed.), *Meaning, Form, and Use in Context: Linguistic Applications*, Georgetown University Roundtable on Languages and Linguistics. Washington, DC, Georgetown UP, 210–28.
Zimmerman, D. 1992. Achieving Context. Openings in Emergency Telephone Calls. In G. Watson and L. R. Seiler (eds), *Text in Context*. Newbury Park, Sage, 35–51.
Zimmerman, D. 1993. Acknowledgment Tokens and Speakership Incipiency Revisited. *Research on Language and Social Interaction*, 26(2), 179–94.
Zorzi, D. 1990. *Parlare insieme*. Bologna, CLUEB.
Zorzi, D. 2002. Il parlato istituzionale: le telefonate al 118. In *N and A*, 11/126.
Zorzi, D. and Monzoni, C. 2004. Le chiamate al 118: pattern ricorrenti nella negoziazione delle informazioni. In G. Bernini, G. Ferrari and M. Pavesi (eds), *Atti del 3° Congresso dell'Associazione Italiana di Linguistica Applicata* (AItLA). Perugia, Guerra, 129–51.

Web resources

Charles Antaki's online introduction to transcription: http://www-staff.lboro.ac.uk/~ssca1/transintro1.htm
Schegloff's online tutorial on transcription: http://www.sscnet.ucla.edu/soc/faculty/schegloff/TranscriptionProject/index.html

Subject Index

account   15, 16, 45, 55, 57–9, 65, 67, 78, 87, 92, 105, 106, 109–12, 116, 123, 124
   see also justification
acknowledgement tokens   19, 21, 73, 79, 101, 114, 169
actual response   71, 73, 89, 93, 103, 111, 116, 123, 130
additional information   3, 47–51, 65, 66, 92
   see also more information
adjacency pair   4, 5, 7, 11–13, 34, 72, 80, 91, 94, 96, 97, 103, 107, 108, 112, 138
agreement   14, 15, 16, 132, 168, 170
   see also disagreement
allocation of turns   9, 10
   see also turn allocation
alternative solutions   51–5, 59, 65–8, 70, 91, 92, 105, 106, 109–12, 117, 123, 124, 130, 139
another turn   35, 47, 69, 93, 107, 114, 115
   see also other turn; different turn; following turn; separate turn
apologising   46, 66, 98, 116, 125
apology   45–7, 58, 60, 65, 78, 109, 110, 113, 114, 125, 126, 130
applied CA   3, 4
architecture   4, 25, 28
ask for clarification   132
asking for confirmation   6, 57, 71, 74, 75, 90, 98
   see also confirmation; confirmation check
asking for more details   48, 53, 71, 78, 79, 98
   see also more details; request for details

business details   139
business of the call   4, 96, 122

C's requests   29, 40, 44, 45, 50, 53, 64, 73, 79, 88, 111, 126, 130, 131
CA   1, 5, 24–9, 140, 142
CA framework   37
CA methodology   29
call centre
   operator   4, 6, 29, 126, 130–2, 136–8
   training   5, 118–37
call centres   113, 118–21, 123, 136, 137
caller leads   96–107
categorisation   27, 37, 76, 92, 110
change of state token   18, 19, 60, 61, 89, 100, 101, 168
clarification request   128–31
closing
   sequences   4, 6, 42, 96, 98, 99, 106
   the call   40
closure relevant   20, 21, 61
combination
   of actions   106, 109, 112, 113
   of strategies   109
common strategies   143
communication strategies   137
complaints   11, 121
complex conversational
   sequences   13
complex formats   5, 77, 108, 113
complex responses   40, 59–61, 113, 139
complex sequences   94
complex structures   72, 93, 102, 106, 110, 116
components   8, 9, 11, 14–19, 97, 101, 126
conditional relevance   12, 14
confirmation   76, 77
confirmation check   71–3, 75, 76, 90, 93, 114, 121
   see also asking for confirmation; confirmation
constituency   4, 5, 7, 23, 29, 30, 35, 47

Subject Index   175

constituents   23, 36, 47, 48, 107, 114, 115
  see also free constituent
constructional component   8, 9
continuative tokens   21, 61
continuers   20, 42, 48, 73, 103, 114
  see also fillers
conversation
  analysis   1, 7, 29
  analysts   3, 8, 21, 27, 28, 30, 36, 37, 142
conversational exchange   35, 36, 99, 121, 141
conversational genre   4, 29, 30, 138, 139
conversational ritual   1, 4, 79, 89
conversational routines   1, 27, 47, 121, 138, 170
conversational strategies   5, 29, 141
conversational studies   3, 25
  see also CA; response strategies; strategies; different strategies
core
  business   2, 123
  response   36, 47, 89, 107, 114–17
correction   16, 17, 58, 101, 171
  see also repair
cross-cultural comparison   5, 65–7, 72, 90, 91, 104–7, 111, 138, 140, 143
cross-cultural differences   31, 138, 144
cross-cultural perspective   29, 30
current activity   12, 98
current speaker   8, 23, 73, 87
customer care   119, 120, 122
customer service   59, 118, 119, 138

DA   25, 27
delay   16, 40, 74, 89, 90, 109, 111, 116, 136
delayed   12, 15, 70, 71, 91, 96, 101, 105, 109, 112, 113, 126, 131
different strategies   110, 123, 143
  see also conversational strategies; strategies
different turns   35, 50, 93, 115
  see also other turn; another turn; following turn; separate turn

disagreement   12, 14–16
  see also agreement
discourse
  analysis   25
  marker   17, 18, 25, 28, 69
discourse and society   27
dispreference   40, 67, 102
  see also preference organisation
dispreferred features   5, 16, 109, 114
dispreferred format   15, 59, 72
dispreferred response   61, 110
dispreferred seconds   14, 15, 68
  see also entries beginning preferred

efficiency of the response   121, 122, 128, 136, 137
elliptical response   98–101
elliptical response token   108
embedded increment   68, 116
ethnography of speaking   25, 28
expand the responses   66, 105, 115, 116, 122
expanded responses   4, 9, 53, 64, 71, 72, 93, 109, 116
expanding the response   94, 112
expansion   7, 9, 13, 23, 35, 36, 47, 51, 64, 65, 69, 75, 79, 87, 92, 93, 104, 105, 107, 109–11, 114–17, 122, 123, 127, 130
  see also extension and increment
explanation   16, 28, 55, 57, 66, 67, 75, 92, 124, 125, 127, 128
extend their response   3, 65, 66, 69, 126, 139
extended   22, 61, 112, 123, 129
  responses   29, 94, 96, 124
  sequences   13
  turn   20, 24
extending the response   55, 61, 65, 66, 124, 139
extension   4, 5, 22, 35, 36, 47, 48, 50, 51, 55, 56, 65–9, 87, 89, 93, 96, 103, 106, 107, 114–16, 123, 129, 142
  see also expansion and increment

face-to-face encounters   1, 32
fillers   109
  see also continuers

following turns 50, 60, 64, 75, 81, 84, 98, 129, 130, 131
  see also another turn; other turn; different turn; separate turn
format of response 5, 36, 39, 40, 44, 66, 67, 75, 77, 91, 92, 108, 110, 113, 116, 129
  see also response format
free constituent 23, 24, 47, 50, 51, 53, 55, 64, 69, 75, 93, 94, 114–16
  see also constituent
free increment 68, 93, 94, 107, 115
  see also increment
free-standing increment 35, 36, 47, 114, 116

grammar in interaction 7, 22
  see also interaction and grammar
grammatical configuration 4, 5, 6, 21, 22, 79, 107, 108, 116, 117
grammatical constituency 5, 7, 29, 30, 35, 47

implications for training 113, 117, 118, 122, 136, 137, 138, 144
inbound call centres 119, 120
  see also outbound call centres
incoming calls 33, 137
incremented responses 93, 114
increments 22–4, 35, 36, 47, 49, 53, 55, 59, 60, 65, 67, 71–95, 105–7, 109, 115–17
  see also expansions and extensions; free increments
insertion sequence 9, 12, 71–3, 75, 77, 79–81, 83, 87–94, 96, 103, 104, 106, 109–12, 114–16, 123, 129
institutional contexts 4, 125, 172
institutional framework 138
institutional roles 4, 29
institutional settings 2–4, 168, 172
institutional talk 3, 4, 6, 30, 137
instruction 119, 122, 123, 125, 126, 141
interaction and grammar 22, 167, 169, 170
  see also grammar in interaction
interactional linguistics 5, 21, 22, 29, 30, 171

interactional organisation 28, 116, 141
interactional sequence 108, 140
interactional sociolinguistics 25, 28
interactional strategies 25, 28, 122, 124
intercultural communication 143
intercultural perspective 30

justification 11, 12, 65, 66, 109
  see also account
justified 55, 91
justifying the response 66

lack
  of service 17, 65, 79, 126
  of training 136
linguistic practices 22, 30

minimal formats 44, 94, 100, 109
  see also simple format; simple response
minimal response token 44, 57, 108
minimal responses 44, 59, 92, 98, 99, 106, 107
minimising 55, 58, 129
mitigated 16, 66, 77, 106
mitigating strategy 66, 126
mitigations 15
more complex formats 5, 77, 108, 113
more complex responses 59, 139
more complex structure 5, 72, 93, 102, 106, 110, 116
more details 34, 48, 53, 71, 78, 80, 90, 91, 98
  see also asking for more details; requests for details
more information 45, 48, 55, 64, 65, 66, 67, 77, 91, 92, 98, 102, 105, 106, 109, 110, 111, 112, 117, 123, 129, 130, 139
multi-unit turn 8, 22, 97, 114

negative response 58, 72, 72, 89, 98, 129, 130
  see also positive response
negotiation 81, 84
news delivery 94, 114, 115

Subject Index   177

news receipt   47, 132
non-satisfaction of the request   12, 39, 126
non-satisfying response   34, 55, 114, 123, 124, 126

offer an alternative item   51, 124
offer of a solution   45, 53, 55, 59, 60, 66, 67, 75, 89, 94, 106, 109 110
oh-receipts   17–19
okays   19–21, 61
opening sequence   1, 4, 30, 139, 166
ordinary conversation   3, 30
organisation of sequences   11, 17, 27, 106, 138
   see also sequence organisation
organisation of the responses   5, 30, 116
other turn   68, 93, 107, 115, 117
   see also another turn; different turn; following turn; separate turn
outbound call centres   119, 120
   see also inbound call centres

participant orientation   139
passive recipiency   19, 20, 85, 87
PCP   8, 23, 48
   see also point of possible completion
phone call   2, 18, 21, 30, 34, 118, 139, 166, 169–73
   see also telephone call
phone openings   2, 144
point of possible completion   22–4, 47, 48, 87, 103
   see also PCP
politeness   78, 99
positive assessments   128, 129
positive response   58, 72, 77, 94, 132
possible completion   22, 47, 48
practical implications   4, 6, 29, 117, 143
practice for call centres   120–2
practices   4, 6, 22, 29, 30, 37, 118, 126, 141, 167, 168
pragmatic completion   9, 23, 87
pragmatic markers   18
pre-closure markers   61
pre-sequence   13, 58, 64, 77
preface   16, 19, 77, 97, 101, 102, 114

prefaced   14, 99, 112, 125, 168
preference   5, 14, 15, 16, 66, 67, 102, 166, 170, 171
   organisation   13, 16
   see also dispreference
preferred character   43
preferred formats   15
preferred second-pair parts   15, 50
preferred seconds   14
preferred turns   15, 170
pre-request for service   2, 4, 5, 13, 32, 40, 50, 118
professional conversation   122
professionalism   122, 123, 125
professionals (in call centre)   120, 121, 125
public service encounters (PSEs)   140, 166

Q–A sequences   100, 102–4
   see also question–answer sequences
qualitative analysis   37, 38, 140
qualitative perspective   36
qualitative studies   140
quantification in CA   36–8, 140, 141, 171
quantifying   37, 141
quantitative analysis   36, 56, 108, 140, 141
quantitative methods   36–8, 138
question–answer sequences   84, 99, 100
   see also Q–A sequences

reason for calling   42, 64, 139
receipt   17–21, 40, 47, 68, 69, 79, 106, 114, 132, 139
receipt token   35, 60, 69, 79, 103
recipiency   19, 20, 23, 24, 87
reformulation   16, 98
remedial strategy   114
remedial work   109
repair   5, 16, 17, 27, 55, 58, 59, 65, 78, 79, 88, 126, 170, 171
   see also correction
repair sequence   58, 71, 89
repaired   64, 126
repairing sequence   58, 89
repeats   43, 46, 64, 77

Subject Index

repetition 35, 42, 43, 47, 51, 53, 57, 59–61, 69, 71, 73, 76, 77, 89, 98, 108, 121, 141, 171
request
  for details 77–9, 83, 90, 92, 114
  for information 4, 11, 28, 30, 32, 37, 50, 51, 118, 121, 122, 123, 129, 132
  preface 97
  for services 40, 114, 116, 143
  see also asking for more details; more details
request–response 4, 50, 73, 80, 91, 96, 104, 107, 108, 138
request–response sequences 4, 12, 29, 34, 35, 37, 94, 95, 106, 108, 118, 137, 143, 144
respond minimally 44, 92
response
  expansion 35, 105, 109, 110, 122, 123
  extended 45, 112
  format 5, 34, 36, 39, 44, 45, 50, 61, 71, 72, 91–3, 105, 108–13, 116, 125, 144
  plus extension 45–70
  to requests 5, 11, 13, 34, 35, 39, 46, 48, 53, 61, 65, 67, 70, 71, 75, 79, 83, 89, 96, 108, 122, 123, 125
  to service requests 145
  strategies 91, 118
  token 17–21, 44, 51, 57, 60, 78, 108, 109, 114
  turns 4, 5, 11, 22, 24, 29, 30, 51, 59, 61, 66, 71, 109, 115, 123
  see also conversational strategies; different strategies; strategies

same turn 17, 35, 47, 49, 53, 55, 58, 68, 69, 75, 93, 107, 114–16, 129
satisfaction 2, 12, 39, 55, 59, 103, 122, 124, 126
  of the request 2, 12, 39
satisfactory response 83
satisfy the request 34, 59
satisfying a customer 34
satisfying response 53, 124
second assessments 14

second-pair part 11–15, 34, 35, 48, 55, 67, 68, 79, 84, 91, 103
separate turn 69, 87, 94, 115, 116
  see also another turn; different turn; following turn; other turn
sequence
  construction 108
  initiated 72, 84, 92, 103, 116, 136
  initiation 80
  organisation 11, 171
  type 143
  see also organisation of sequences
sequential deployment 24, 96, 104
sequential development 71, 106
sequential exchange 7
sequential moves 137
sequential order 121, 123, 124
sequential organisation 27, 138
sequential patterns 2
sequential position 107
sequential response format 91
sequential rules 144
sequential shape 4
service
  encounters 1–7, 13, 17, 28, 29, 37, 39, 67, 71, 90, 104, 105, 107–10, 114, 116, 118, 122, 123, 136, 138, 140, 144, 166, 167, 172
  offer 99, 122
  providers 1, 4, 6, 29, 30, 37, 93, 118–20, 122, 124, 132, 137, 142, 143
  requested 3, 45, 50, 55, 66, 67, 72, 92, 94, 98, 102, 105, 109, 124, 139
  see also telephone service encounters
simple format 6, 42, 80, 94, 96, 103, 112, 113
  see also minimal format
simple response 3, 39, 91, 93, 94, 96, 103, 104, 109, 110, 111, 113, 115, 116
  format 5, 39–44, 72, 116
simple sentences 116
simple turn 71, 72, 113
simply formulated responses 44
small businesses 30, 138
small companies 118

Subject Index 179

small services 120, 137
social interaction 5, 7, 28, 142
social roles 2, 4, 29
solicitations 98, 101, 104–7, 109
staff 29, 30, 33, 48, 93, 124, 126
standardisation 121, 122, 136
story prefaces 77
story-telling 8, 13
see also telling a story
strategies 4, 5, 25, 28, 29, 47, 59, 66, 67, 75, 77, 90–2, 109, 114, 117, 119–26, 129, 139–45
see also conversational strategies; response strategies; different strategies
strategies of response 110, 129
see also response strategies
structural complexity 39, 71, 72, 93, 108
structural formulation 67
structure of the conversation 3, 100
syntactic complete utterances 9
syntactic completions 8, 9
syntactic constructions 30, 47
syntactic developments 115
syntactic units 8, 114
syntactically dependent 56
syntactically independent 35, 36, 64, 69, 75, 79, 84, 89, 103, 107, 114
syntactically organised 116
syntactically tied 87, 107

talk-in-interaction 7, 23, 24, 37, 138, 141–3, 145
TCU 8, 9, 17, 20, 24, 28, 69, 97, 115, 142
see also turn constructional units
telephone calls 1, 30
telephone customer care 122

telephone service encounters 29, 108, 110, 114, 116, 118, 122, 137, 138, 144
telephone transaction 137
telling a story 83
see also story-telling
training 5, 29, 59, 65, 83, 93, 118, 120–6, 136
of assistants 138
of call centre operators 6
of service providers 4
training personnel 113, 123
transition relevance place 8, 21
trouble source 16, 17, 89
TRP 8, 9, 21, 24, 69
see also transition relevance place
turn
allocation 8, 9
beginner 129
beginnings 17, 18
construction 22, 34
constructional component 8, 9
contructional units 22, 50
design 14, 17, 29
-taking 7, 11, 17, 20, 27, 34
-taking system 4, 5, 7, 29, 34
see also TCU
transition 51, 69
unit 8, 9, 23, 35, 47, 48, 69, 72, 75, 79, 83, 84, 87, 114, 115
type
of encounter 143
of service 31, 32, 65, 77, 142
of turn 5, 23, 35, 36, 65, 101

unattached NPs 23, 24, 35, 47
understanding C's needs 126–37

weak acknowledgement token 20, 73, 87
whose voice has to speak? 124, 125

Author Index

Antaki 173
Armstrong 36, 166
Aston 1, 3, 26, 68, 109, 114, 140, 142, 166
Atkinson 3, 13, 15, 33, 166, 168, 170
Auer 12, 22, 166

Barnes 36, 166
Beach 19, 21, 61, 166
Bercelli 2, 139, 166
Bergman 126, 139, 166
Billig 27, 166
Bilmes 13, 166
Bonu 13, 166
Bowles 2, 90, 166, 170, 172
Boyle 13, 166
Brodine 1, 3, 68, 109, 114, 142, 166

Cameron 121, 125, 136, 137, 166
Church 13, 166
Clark 11, 166
Colamussi 2, 141, 167
Couper-Kuhlen 21, 22, 167, 171

De Stefani 142, 167
Doany 13, 168
Drew 3, 4, 13, 15, 36, 125, 137, 141, 142, 166, 167
Drummond 19, 20, 167
Du Bois 9, 167

Eggins 25, 167

Finch 120, 167
Ford 9, 22–4, 28, 35, 47, 87, 167, 168
Fox 22, 167, 168
Francis 3, 168

Gardner 19, 20, 21, 167
Gavioli 8, 142, 167, 168, 172
Gilbert 27, 168
Goffmann 140, 168
Goodwin 22, 23, 26, 168, 171

Greatbatch 3, 168
Gumperz 25, 168, 169
Guthrie 19, 21, 168

Halmari 2, 139, 168
Hart 34, 168
Heritage 3, 4, 13, 14, 17–19, 26, 33, 36, 61, 100, 101, 125, 137, 142, 166–8, 170
Hester 3, 168
Hopper 2, 13, 19, 20, 107, 167, 168
Hutchby 2, 33, 37, 142, 168
Hymes 25, 169

Jefferson 12, 19, 20, 33, 87, 169, 170, 171
Joyce 120, 169
Jucker 18, 169

Kasper 126, 166, 170
Koke 120, 172

Lee 1, 3, 34, 142, 169, 170
Leidner 120, 169
Leonardi 3, 167, 169, 172
Lerner 17, 50, 169
Lesser 36, 169
Levinson 12–16, 37, 68, 109, 114, 169
Lieflander-Koistinen 2, 169
Lindstrom 22, 69, 169
Luke 3, 139, 169, 171, 172

McHoul 3, 168, 169, 172
Mehan 3, 169
Menzler Trott 120, 169
Mey 13, 169
Mondada 37, 169
Monzoni 3, 142, 169, 173
Mulkay 27, 168

Ochs 22, 167, 169, 170, 171
Olshtain 126, 170
Ono 22, 47, 170

Pallotti 2, 26, 30, 37, 90, 139, 141, 166–8, 170
Pavlidou 3, 139, 169–72
Pomerantz 14, 15, 102, 109, 170
Psathas 11, 118, 125, 142, 168, 170

Roberts 3, 170
Ronchi 120, 170

Sacks 3, 7, 8, 9, 11, 14, 17, 21, 24, 98, 142, 170, 171
Sarangi 3, 170
Schegloff 1, 11–17, 20–3, 25, 27, 28, 33, 35, 36, 38, 47, 55, 98, 109, 138, 140, 142, 144, 167, 169, 170, 171, 173
Schiffrin 18, 171, 173
Seedhouse 25, 37, 167, 170–2
Selting 8, 9, 20–2, 28, 142, 167, 171
Sidnell 142, 171
Silverman 3, 171
Slade 25, 167
Steffen 120, 172
Stivers 142, 171

Svennevig 46, 47, 171
Szymanski 1, 3, 34, 142, 172

Tannen 51, 170, 171
Ten Have 3, 4, 33, 34, 37, 125, 142, 144, 171, 172
Thieme 120, 172
Thompson 9, 22, 28, 47, 167–71
Thuene 2, 3, 139, 167, 169, 172
Trosborg 125, 172

Varcasia 2, 90, 124, 126, 170, 172
Vinkhuyzen 1, 3, 34, 142, 172

Wagner 22, 167
Wakin 139, 172
Whalen 3, 142, 172
Wiencke 120, 172
Wooffitt 2, 27, 33, 37, 142, 168, 172

Yule 13, 172

Zimmerman 3, 28, 139, 142, 172, 173
Ziv 18, 169
Zorzi 1, 3, 68, 109, 114, 142, 169, 173

The manufacturer's authorised representative in the EU is Springer Nature Customer Service Centre GmbH, Europaplatz 3, 69115 Heidelberg, Germany. If you have any concerns regarding our products, please contact ProductSafety@springernature.com

Printed and bound by CPI Group (UK) Ltd, Croydon, CR0 4YY

23/03/2026

02076449-0013